FABRICATION OF PALLADIUM AND PLATINUM NANOPARTICLES DECORATED REDUCED GRAPHENE OXIDE AS A HIGHLY ACTIVE AND STABLE ELECTROCATALYST

A Thesis presented to the Faculty of the Graduate School of the University of Missouri-Columbia In Partial Fulfillment of the Requirements for the Degree Master of Science

Nada Sami Ahmed Al – Janabi

CONTENTS

ACKNOWLEDGMENTS

I would like to express all my appreciation to my advisor Professor John Gahl, and SKINR group. I also want to sincerely thank Dr. Orchideh Azizi. Everyone involved in my project gave precious and tireless advice, encouragement and support throughout my graduate studies. They have allowed me to grow as a research scientist. Also I want to give my sincere thanks to Dr. Kattesh Katti's group and Dr. Shubhra Gangopadhyay's group, who supported my research.

I also want to extend my thanks to Dr. Paul Chan in the Department of Chemical Engineering for being on my graduation thesis committee. His guidance and enthusiasm are greatly appreciated.
In addition, I will give my gratitude to my sponsorship HCED, whose gracious help meant a lot to me.
Thank you to all my relatives and friends who are either in U.S.A or in my country, Iraq.

Last but not least, I would like to convey a special thanks to my beloved parents and my dear sisters, whose moral encouragement and support helped me realize my master's degree goal.

Columbia, July 2016

Nada Sami

ABSTRACT

In this work, we developed new types of flexible electrodes based on chemical and electrochemical techniques by decorated reduced graphene oxide with platinum (Pt) and palladium (Pd) nanoparticles. In order to obtain reduced graphene oxide sheets, we used HI acid and EGCG as chemical reduction agents. The electrical conductivity was higher for a RGO sheet which was obtained from HI acid treatment than the samples using EGCG. The HI acid treatment exhibited a sheet with very good flexibility and a much higher tensile strength than the samples with EGCG. Chemical deposition method showed the ability of HI acid and EGCG to reduce platinum salt (Pt salt) to platinum nanoparticles (Pt NPs). All the chemical deposition samples with EGCG treatment showed a good particle distribution with a small size and less aggregation. Unlike the samples that were treated with HI acid only. One-step electrochemical deposition from different types of electrolytes was demonstrated. Morphological and structural characterizations showed that Pd NPs can be efficiently, with a very high stability and activity, decorated electrochemically on RGO from Pd NP suspension electrolyte. In addition, electrochemical deposition from (PdII complex) resulted in a very high density of Pd decorated on the surface of RGO sheet. In conclusion, this research demonstrated a variety of desirable electrocatalysis properties including excellent electrocatalytic activity, small Pd nanoparticles dispersed evenly on the surface of the electrode, and very high stability with exceptional flexibility.

CHAPTER 1

INTRODUCTION

Most catalysts consist of nanometer-sized particles dispersed on a high-surface area support [1]. Previously, graphene oxide (GO) has been studied as a material of broad interest and potential because of its extraordinary properties, such as high surface area, good thermal conductivity, outstanding electrical conductivity, and optical transparency [2]. Many studies, including ours, have reduced graphene oxide (GO) by using chemical reduction agents to remove the oxygen-containing groups in order to obtain a reduced graphene oxide sheet (RGO) with high conductivity. Many reduction agents like HI, HBr, and HH acids have been used in an attempt to increase efficiency of reduced GO and improve the final properties [3]. Because HI acid is one of the strongest reduction agents, we used this acid to modify a flexible and high conductive reduced GO electrode.

In addition to HI acid, various authors recently reported green tea as a green reduction agent. Tea polyphenols (TPs) is an extract of green tea mainly consisting of four major catechins: epigallocatechin gallate (EGCG), epigallocatechin (EGC),
epicatechin gallate (ECG), and epicatechin (EC) [4]. The EGCG makes up about
50 – 80 wt % of the total catechins in green tea. Epigallocatechin gallate (EGCG) acts as an efficient reducing agent and stabilizer simultaneously [4]. Both of the previous chemical reduction agents, HI acid and EGCG, were used to produce on a large scale and low cost RGO sheets (substrate).

RGO can be made as a high-performance flexible electrode for electrocatalysis production by improving the electrical conductivity, and mechanical strength with the reduction process. Reduction also provides more nucleation sites for the metal nanoparticles.

3

The chemical and electrochemical depositions of precious metals play an important role in catalyst production [5]. There are many flexible catalysis types synthesized by using chemical deposition. The electrochemical deposition technique is the most effective, with a high efficiency, approach to prepare metal nanoparticles and then deposit them on the surface of the substrate with easy shape control which allows for different sizes, composition, and morphology. Platinum groups have long been the most interesting material for many research studies concerning the production of catalysts. In this work,

a one-step chemical deposition was used to modify many electrode types after the deposition of platinum nanoparticles (Pt NPs) from platinum salt (Pt salt) on the surface of different reduced GO sheets.

In this study, the aim is to reduce graphene oxide using different procedures and then fabricate two kinds of RGO – Pd catalyst by taking the advantages of Pd nanoparticles electrodeposition (Pd NPs) from different types of electrolytes. These modified

RGO – Pd compositions can be used as a catalyst with high activity, high stability and distinctive flexibility for chemical reactions, such as Hydrogen/Oxygen recombiner and as an electro-catalyst for electrochemical reactions such as water electrolysis, methanol oxidation, and so on.

CHAPTER 2

MATERIALS AND METHODS

1-   Synthesis GO:

Preparing GO sheet electrode: Graphene oxide has been synthesized from oxidizing graphene to GO by the modified "Hummers and Offenman's" method [6]. The aqueous suspension from the GO synthesis was then ultrasonicated and dried to form a sheet of GO.

2-   Reduction of GO:

Preparation of GO – (EGCG) sheet electrode: required amount of EGCG was first dissolved in 10 mL of MQ – water and treated with sonication for 30 min. Then, the GO sheet was immersed into aqueous solution of EGCG for two days to obtain GO – (EGCG) sheet. The GO – EGCG was washed with MQ – water for several times and dried at room temperature.

Preparation of (rGO – EGCG) and (rGO – 2EGCG) modified sheet electrodes: EGCG (12.5 wt % relative to GO) was added to a GO suspension for (rGO – EGCG) sample and (25 wt % relative to GO) for (rGO – 2EGCG) sample. The mixture was sonicated for 1 hour; then dried at 40 °C.

Preparation of RGO modified sheet electrode: The GO sheet was directly immersed into hydroiodic acid (HI) solution in a sealed cuvette at room temperature for 1 hour. Then, the RGO sheet was washed with MQ – water for several times and dried also at room temperature to obtain the final sheet electrode [11].

3- Electroless deposition (chemical deposition):

Preparation of GO – (Pt) and GO – (EGCG Pt) sheet electrodes: To begin with, 8.75 mg Pt salt was dissolved in 25 mL of MQ – water and sonicated for 30 min. In parallel, 2.27 mg EGCG was dissolved in 10 mL of MQ – water. The first GO sheet was immersed in the solution of Pt salt for 5 h to obtain GO – (Pt) sheet and the other was immersed in (50 % of EGCG solution + 50 % of Pt salt solution) to obtain GO – (EGCG Pt) sheet. After that, these two sheets were washed several times and dried at room temperature.

Preparation of (rGO – EGCG) – Pt and (rGO – 2EGCG) – Pt sheet electrodes: The (rGO – EGCG) and (rGO – 2EGCG) sheets immersed in the solution of Pt salt for 5 h, then these sheets were washed several times with

MQ – water and dried at room temperature.

Preparation of RGO – (Pt) and RGO – (EGCG Pt) sheet electrodes: To prepare RGO – (Pt) we immersed two RGO sheets in the solution of Pt salt, one for 5 h and the other for 24 h. RGO – (EGCG Pt) was prepared by immersing the RGO sheet in the solution of (EGCG + Pt salt) for 24 h. After the surface treatment, we washed these samples several times with MQ – water and dried them at room temperature.

4-   Electrodeposition (electrochemical deposition):

Preparation of RGO – Ed Pd sheet electrode: Palladium was deposited under galvanostatic conditions with a current density of 6.2 mA cm-2 on the surface of RGO at room temperature. The time for electrodeposition process was 1 hour and 5 hours. The electrodeposition solution was acidic electrolyte bath containing

(5 gm/L PdCl2 + 0.1 M HCl).

Preparation of RGO – Ed Pd NP sheet electrode: Palladium was deposited under galvanostatic conditions at room temperature with a current density of 6.2 mA cm-2 on the surface of RGO. The electrodeposition times were 24 hour and 48 hours. The electrodeposition solution was Pd NP suspension (alkaline bath) electrolyte bath containing (Pd NP + 0.1 M LiOH). Note the Pd NP suspension was prepared from Pd nanoparticles suspension with EGCG in professor Katti's lab.

5-   Characterization and electrochemical measurement:

Scanning Electron Microscopy (SEM) images were obtained using FEI Quanta 600 FEG Environmental Scanning Electron Microscope (ESEM), equipped to perform elemental chemical composition analysis by Energy Dispersive X-ray Spectrometry (EDS/EDX). All electrochemical studies including cyclic voltammetric (CV) and chronopotentiometry (CP) experiments were performed with a Single Potentiostats / Galvanostats SP-300 electrochemical workstation (Bio Logic since instruments). During the electrochemical measurements, the working electrode was a contacted reduced graphene oxide (RGO) sheet electrode set in cell device, and the auxiliary and reference electrodes were Pt wire, and Ag/AgCl, respectively. Cyclic voltammetry CV analysis was used to investigate the electrochemical behaviors of all sample electrodes in 0.1 M H2SO4 solutions at different Scan Rates of (mV/s).

CHAPTER 3

RESULTS AND DISCUSSION

3.1- Preparation and Characterization of the Freestanding Reduced Graphene Oxide (RGO) Electrode:

There are a number of routes for the reduction of GO, such as chemical reduction, thermal reduction, electrochemical reduction and photocatalytic reduction [7]. Reduction by chemical reagents is based on their chemical reactions with GO. On the other hand, the reduction can make a great change in the microstructure and properties of GO by removing functional groups like the oxygen containing group. This can result in a highly conductive sheet electrode. The chemical reduction method is the cheaper and more easily available way for the mass production of GO with high efficiency, especially when compared with the other methods of reduction like thermal reduction. In addition, the requirement for equipment and laboratory environment is not as critical as that for thermal annealing treatment; for example, the chemical reduction can be performed at room temperature, which makes chemical reduction one of the favorite and most effective methods of GO reduction [9].

Synthesis and characterization of reduced graphene oxide sheet by EGCG (polyphenol); (GO – (EGCG)), (rGO – EGCG) and (rGO – 2EGCG) sheets:

EGCG was employed as a reducing agent and stabilizer for GO [4, 8]. The color after making soluble EGCG in graphene oxide changed from brownish–yellow to black because of the reaction between GO and EGCG. Corresponding EDX analysis was conducted to determine the atomic percent (at %) of the elements present in the samples in this research. The results

7

from EDX spectra are presented in Table 1. EDX was performed to identify oxygen-containing functional groups, the O atoms occurred by the forms of O-C=O, C-OH, and C=O can clearly identify the reduction ability of EGCG to GO from surface atomic C/O ratio. For the GO sample, the intensity of the C/O ratio was ~ 1.6 which increased to ~ 3.1 after adding EGCG (12.5 wt % of graphene oxide weight percentage). After doubling the amount of EGCG in the (rGO – EGCG) sheet, the C/O ratio was raised to ~ 6.3. As shown in Table 1, the effect of EGCG increased by twice after doubling the amount of EGCG for the (rGO – 2EGCG) sheet (25 wt % of graphene oxide weight percentage). The two ways of reducing GO by EGCG exhibit clearly different conductivities. Immersing GO sheet in EGCG solution to obtain a (GO – (EGCG)) sheet results in only a C/O ratio ~ 2.7 after two days, which improves the significance of mixing EGCG in GO suspension. These results are possibly due to a non-uniform size distribution and particle aggregation for EGCG in solution. In contrast, the sonication of EGCG with GO suspension could contribute in a more uniform distribution for EGCG and give more effective reduction for GO. The EGCG shows a distinct capability to reduce GO because most of the phenols on the gallic acid units are converted to galloyl-derived orthoquinone [4].

Note that the EGCG may also contribute to the oxygen groups [8]. In SEM Figures 1 and 2, there is no significant morphology for distributing EGCG in both previous ways. Although EGCG contributed in reducing the oxygen-containing groups, there are still some oxygen groups. Additionally, the efficient reduction and conductivity is not enough for using any previous sheets as an electrode sheet.

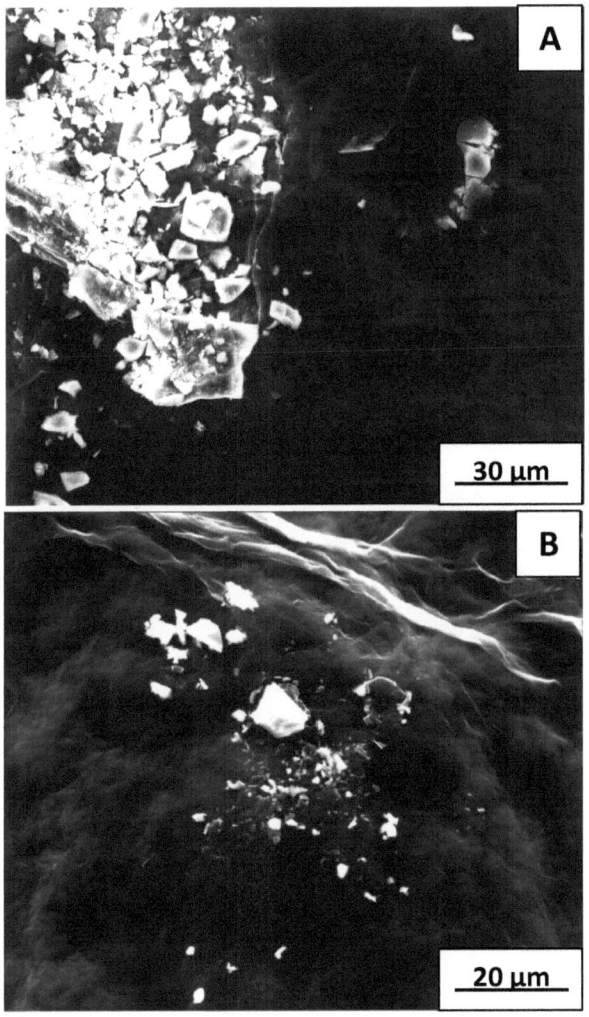

Figure 1. SEM images illustrated EGCG groups in A) rGO – EGCG sheet, B) GO –

(EGCG), sheets.

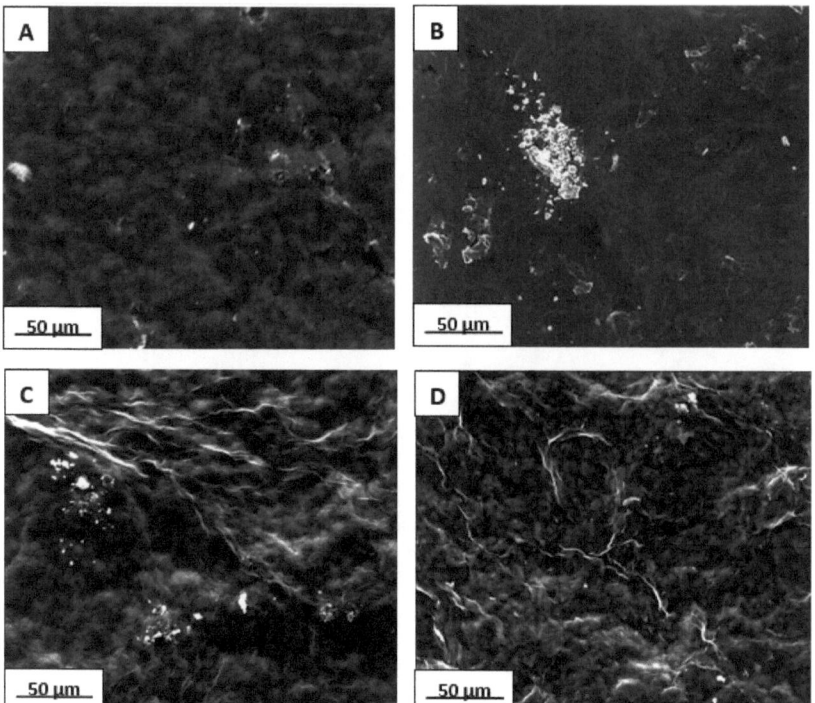

Figure 2. SEM images for A) GO, B) GO – (EGCG), C) rGO – EGCG, D) rGO – 2EGCG, sheets.

Synthesis and characterization of reduced graphene oxide sheet by HI acid; (RGO) sheet: In order to modify a highly conductive RGO electrode, we tried to efficiently remove functional groups like epoxy and hydroxyl, which are the key for the reduction of GO. Halogenation agents, including concentrated hydroiodic acid (HI acid), are strong acidic reduction agents [10]. The resulting RGO sheet from immersing a GO sheet in to HI acid shows a higher electrical conductivity and C/O atomic ratio than other sheets that were reduced by EGCG. The surface atomic C/O ratio is ~ 12.5 on a HI reduced GO sheet; (Table 1).

The value of C/O ratio that was achieved by using HI is much higher than the reduction by EGCG. The optical observation showed the change in color of the GO sheet from black and lackluster to shining metallic luster after reduction by HI, indicating a reduced GO sheet. Reduction by HI acid created a RGO sheet with high electrical conductivity, strong tensile strength, light weight and excellent mechanical flexibility.

Synthesis and characterization of reduced (rGO – 2EGCG) sheet by HI acid; R(rGO – 2EGCG) – HI sheet: In trying to obtain a higher conductivity electrode sheet, the (rGO – 2EGCG) sheet was reduced again in HI acid as a R(rGO – 2EGCG) – HI sheet. The SEM in Figure 3 shows a more shining surface for R(rGO – 2EGCG) – HI because of HI acid reduction. The C/O ratio after this reduction is ~ 10.26 in (Table 1), which is less than the reduction of GO by HI acid. The reduction reaction for GO on the surface of the (rGO – 2EGCG) sheet decreased under the acidic conditions of HI acid, leading to the release of free EGCG [12], which may be the reason for decreased reduction. Even while improving the tensile strength and flexibility of the R(rGO – 2EGCG) – HI sheet, it did not have that quality compared with the RGO sheet for use as an electrode sheet.

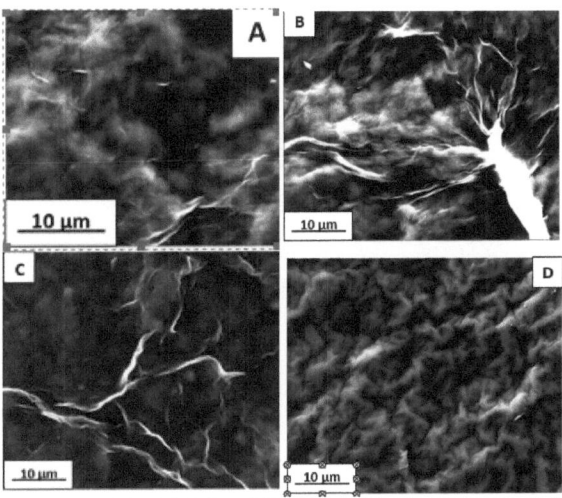

Figure 3. SEM images A and B) R(rGO – 2EGCG) – HI, C) (rGO – 2EGCG), D) RGO, sheets.

Samples	C% Atomic	O% Atomic	Ratio of Carbon to Oxygen
GO sheet	58.94	36.25	~ 1.6
rGO – EGCG sheet	74.04	23.96	~ 3.1
rGO – 2EGCG sheet	83.79	13.28	~ 6.3
GO – (EGCG) sheet	73.27	26.73	~ 2.7
RGO sheet	83.98	6.7	~12.5
R(rGO – 2EGCG) – HI sheet	71.23	6.94	~10.26

Table 1. EDX elementals analysis of GO, rGO – EGCG, rGO – 2EGCG, RGO, GO – (EGCG), R(rGO – 2EGCG) – HI, sheets.

Comparing the previous experimental work with HI acid and EGCG reduction agents, the electrical conductivity for reduction by HI acid is the best for high conductivity of electrode sheet. The reason for low conductivity with EGCG reduction is related to the heavily absorbed EGCG molecules on the graphene surface [8]. Besides, the reduction with HI acid will change physical and mechanical properties of GO. RGO sheet has a very good flexibility with a much higher tensile strength than the samples with the presence of EGCG. In general, all the previous work has proved that the quality of HI acid as a reduction agent for graphene oxide is higher and more effective than EGCG in producing flexible conductors and energy–storage devices.

Electrochemical characterizations of GO and reduced GO sheet electrodes:

CVs of GO vs. (rGO – 2EGCG) sheets: Figure 4 shows the cyclic voltammograms recorded on GO and (rGO – 2EGCG) sheets in 0.1M $H2SO4$ electrolyte with the scan rate of 100mV/s. The current density is significantly expanded and much higher for (rGO – 2EGCG) in comparison to the GO sheet in the whole range of potentials. These results show that EGCG effectively reduced the functional group of GO and as a result, the conductivity, specific surface area and double layer capacitance of the (rGO – 2EGCG) sheet are higher in comparison to the original GO sheet.

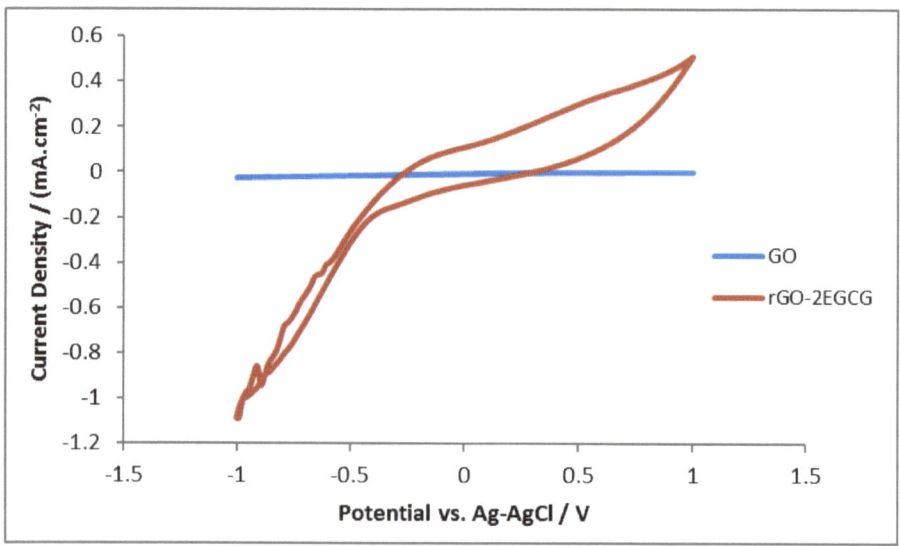

Figure 4. CVs for GO vs. (rGO – 2EGCG) sheets in 0.1 M H2SO4 solutions at a Scan Rates of 100 (mV/s).

CVs of (rGO – 2EGCG) vs. RGO sheets:

Figure 5 shows a cyclic voltammograms of (rGO-2EGCG) and GO reduced in the HI acid for 1 hour RGO in 0.1M H2SO4 solution. The results reveal that the current density in the RGO electrode is much higher than the (rGO – 2EGCG) which is related to improvement in the electrical conductivity behavior of GO treated with HI acid and the formation of more nucleation sites on the surface of an RGO electrode sheet.

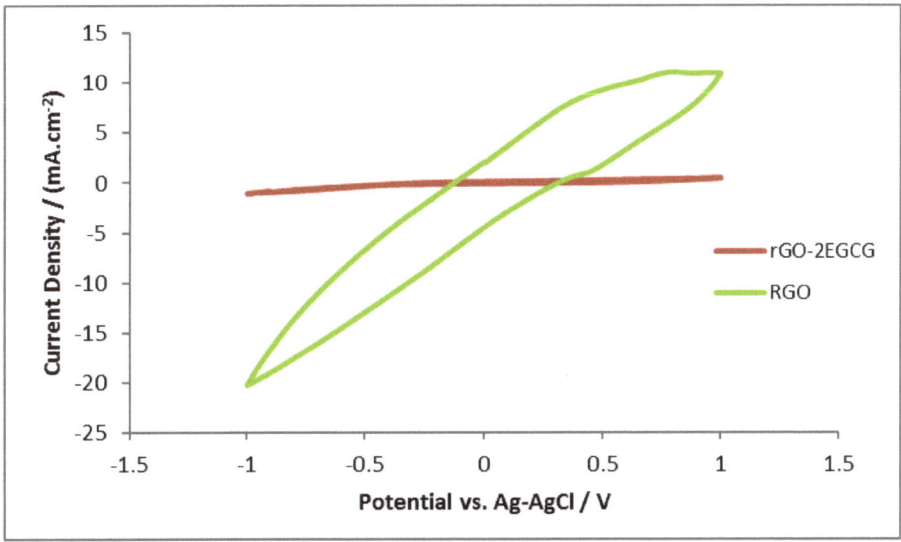

Figure 5. CVs for RGO vs. (rGO – 2EGCG) sheets in 0.1 M H2SO4 solutions at a Scan Rates of 100 (mV/s).

CVs of GO vs. (rGO – 2EGCG) and RGO sheets:

Figure 6 displays the cyclic voltammograms obtained on GO, (rGO – 2EGCG) and RGO sheets in 0.1M H2SO4 solution at a scan rate of 100 mV/s. Comparison between the voltammograms shows that the current density obtained in GO at the whole potential range is much lower than the two other electrodes which is related to high resistivity of GO. The slightly higher current density obtained in GO is treated with EGCG. The highest current density obtained in GO treated by HI acid shows that HI acid is more effective in reducing GO functional groups and provides a faster electron transfer rate and a larger effective surface area than EGCG. The higher current density in the RGO electrode is related to improvement in the electrical conductivity behavior of GO treated with HI acid and formation of more nucleation sites on the surface of an RGO sheet electrode. These results are in good agreement with C/O calculated using EDX data.

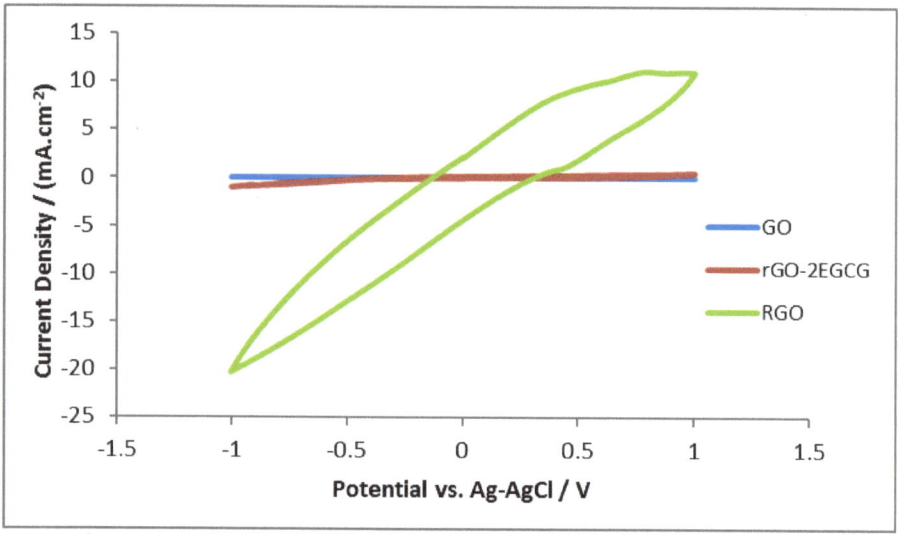

Figure 6. CVs for GO vs. RGO and (rGO – 2EGCG) sheets in 0.1 M
H2SO4 solutions at a Scan Rates of 100 (mV/s).

3.2- Chemical deposition and characterization of forming graphene – Pt nanoparticles composites:

"Electroless metal deposition" (ED) is an auto-catalytic process of depositing a metal in the absence of an external source of electric current. A chemical reducing agent reduces a metallic salt onto specific sites of a catalytic surface which can either be an active substrate or an inert substrate seeded like a conductive, non-conductive, or semi-conductive substrate with a catalytically active metal [13, 5]. Electrolessly deposited platinum has been adopted in the plating industry for many applications, especially since the conventional ED method can produce highly dispersed Pt nanoparticles (Pt NPs) with small particle sizes [13]. The solutions for electroless plating in this research essentially contain EGCG and HI acid (reduction agents) and the source of the metal to be deposited. This creates surface functional groups for metal nanoparticles deposition on the surface of different electrode sheets.

Deposition of Pt nanoparticles on GO sheet from Pt salt (GO – (Pt)) sheet:

Figure 7 shows the SEM images of GO sheet and GO sheet treated with Pt salt solution for 24 h. The SEM and EDX results did not record the formation of Pt nanoparticles (Pt NPs) on the surface of the GO sheet. This is evidence for the need of a reduction agent to deposit Pt NPs from Pt salt on the surface of the GO sheet electrode.

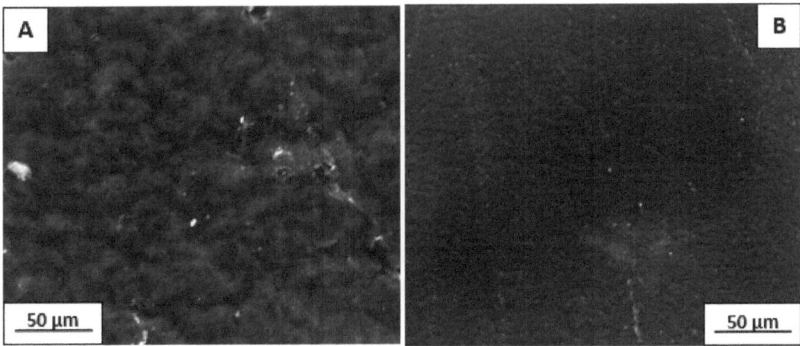

Figure 7. SEM images A) GO, B) GO – (Pt), sheets.

Electroless deposition of Pt nanoparticles from Pt salt by using EGCG as a reduction agent; (GO – (EGCG Pt)) sheet: Previously, the EGCG has been reported as an active reduction agent in the reduction of a GO sheet. Subsequently, the deposition of Pt nanoparticles on the GO sheet was achieved by the assistance of EGCG as a reduction agent. After treating the surface of the GO sheet with the solution of Pt salt with an appropriate amount of EGCG, the $PtCl_4^{2-}$ was partially reduced to Pt NPs and deposited on the surface of GO sheet. Figure 8 shows the SEM image of the GO sheet treated by EGCG and the Pt salt solution for 24 hours. Despite a 24 h treatment of the GO sheet in a mixture of Pt salt and EGCG solution, the EDX recorded a very low loading of Pt nanoparticles (~1.7 wt%) with aggregation of EGCG on the surface of the GO sheet, with the exception of the large EGCG particles as shown in Figure 8.

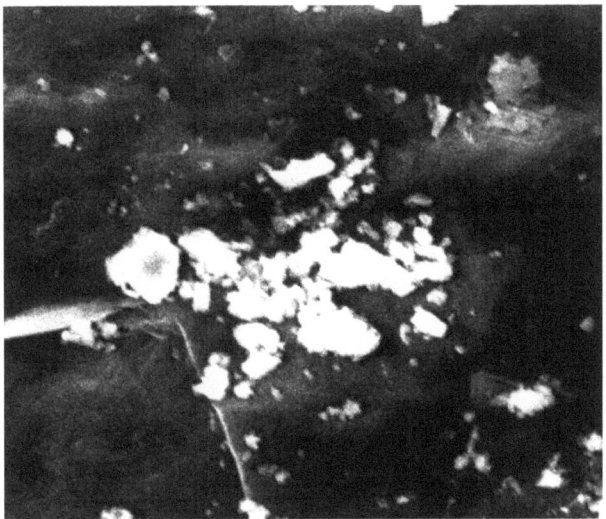

Figure 8. SEM image of the GO – (EGCG Pt) sheet.

Electroless deposition of Pt nanoparticles on the (rGO – EGCG & rGO – 2EGCG) sheets from Pt salt; (rGO – EGCG) – Pt and (rGO – 2EGCG) – Pt sheets: From treatment the (rGO – EGCG) and (rGO – 2EGCG) sheet with Pt salt solution it was noted that the loading of Pt NPs was increased (~9.9 wt% Pt) after treating the sheet of (rGO – EGCG) with Pt salt solution, and it doubled to (~20.03 wt% Pt) after treating a (rGO – 2EGCG) sheet in Pt salt solution. In this case, EGCG acts as a reduction agent for GO and Pt salt at the same time. The SEM images in Figures 9 & 10 show the small size of Pt NPs with good dispersal on the surface of the sheet electrodes. On the other hand, because of the good sonication for the aqueous suspension of GO and EGCG to obtain (rGO – EGCGs) sheets, the EGCG groups became smaller and had less aggregation. That leads to more deposition of Pt NPs with a wide dispersal on the surface of the sheets.

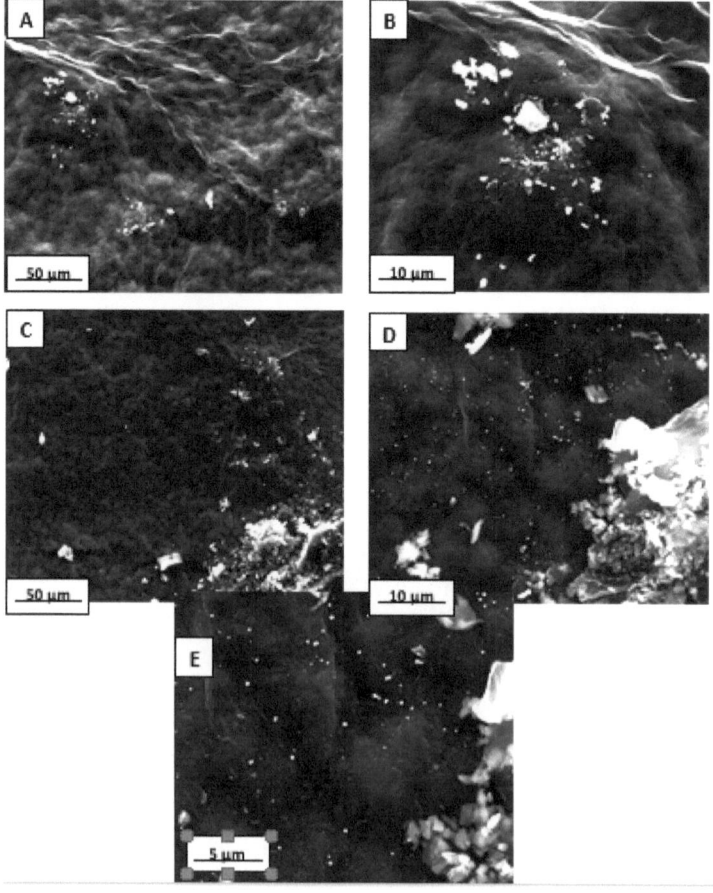

Figure 9. SEM images of A, B) (rGO – EGCG), C, D, E) (rGO – EGCG) – Pt, sheets.

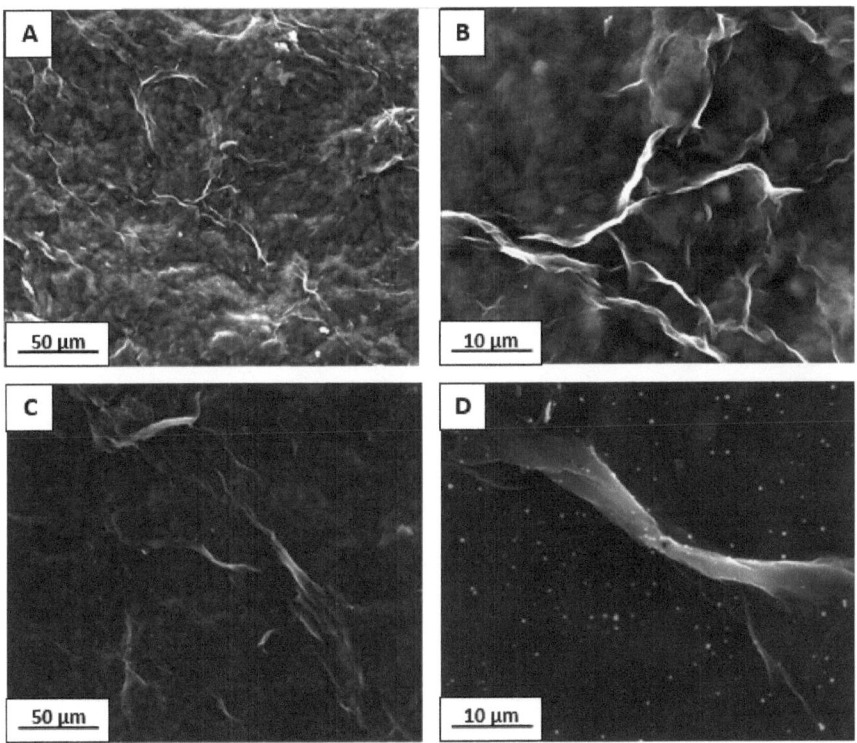

Figure 10. SEM images of A, B) (rGO – 2EGCG), C, D) (rGO – 2EGCG) – Pt, sheets.

Electroless deposition of Pt from Pt salt by using hydroiodic acid (HI acid) as the reducing agent; (RGO – (Pt)) sheet:

RGO sheets have been treated with Pt salt for two time periods, 5 and 24 hours, to obtain RGO – (Pt) sheet. Figure 11 shows the SEM images of RGO sheet treated with the different reduction periods. Both reduction periods reveal the formation of Pt NPs on the surface of the sheets. From EDX analysis, the Pt NPs loading increased from (~3.49 wt %) with 5 hours treatment to (~34.38 wt % Pt) for 24 hours. This means the RGO – (Pt) has a higher loading of Pt NPs than (rGO – 2EGCG) – Pt sheet.

Interestingly, the RGO – (Pt) samples show that HI acid displays a somewhat reductive behavior for Pt salt. This reduction reaction apparently proceeds because of the redox potential of PtCl42-/Pt [+ 0.775 V vs. Standard Hydrogen Electrode (SHE)] is more positive compared to the redox potential of hydrogen iodide acid [+ 0.54 V vs. SHE]. That leads to the chemical reduction of PtCl4-2 by HI acid and will cause the formation of Pt NPs on the surface. The HI acid which remained inside of the RGO sheet reduced Pt salt on the surface of RGO sheet. The SEM images in Figure 11 shows highly aggregated Pt NPs with a high density. After the 24-hour period, there was an increase in the particle size.

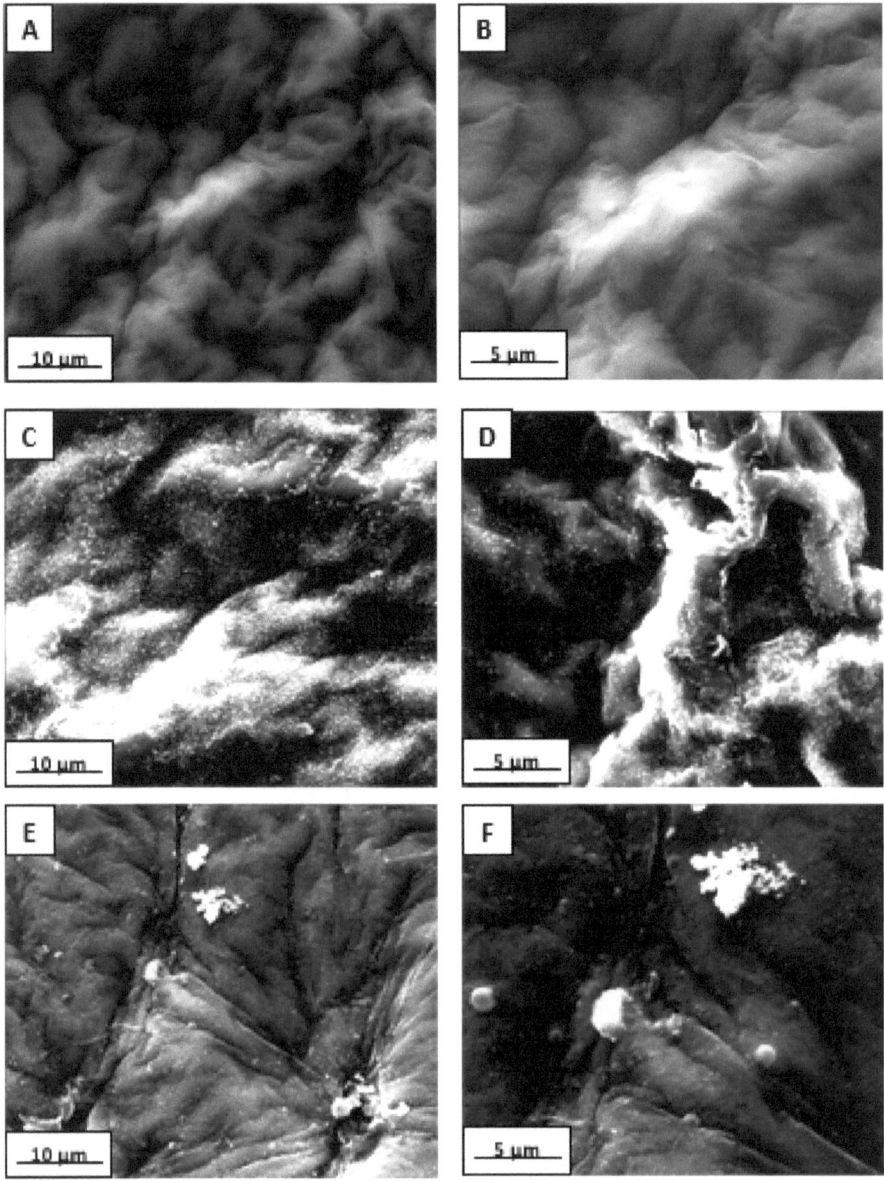

Figure 11. SEM images of A, B) RGO, C, D) RGO – (Pt) for 5 hours, E, F) RGO – (Pt) for 24 hours, sheets.

Electroless deposition of Pt nanoparticles from Pt salt by using HI acid and EGCG at the same time; RGO – (EGCG Pt) sheet: To obtain more understanding, we treated the RGO sheet for 24 hours with a solution of (EGCG and Pt salt). The redox potential of PtCl42-/ Pt [+ 0.775 V vs. (SHE)] is significantly more positive as compared to the redox potential of EGCG [+ 0.42 V vs. SHE] [18]. This results in a thermodynamically feasable redox couple of PtCl42-/ EGCG, leading to the reduction of PtCl42- by EGCG to form Pt NPs with (~17.92 wt% Pt) loading. This result is lower than almost all previous experiments. Because a HI acid environment leads to the release of free EGCG, this causes a decrease in the reduction ability for both agents. The deposition of PtCl42- with each agent alone gives results better than using them together at the same time. The results depicted in the SEM in Figure 12 show less aggregated particles and more uniform dispersal of smaller particles.

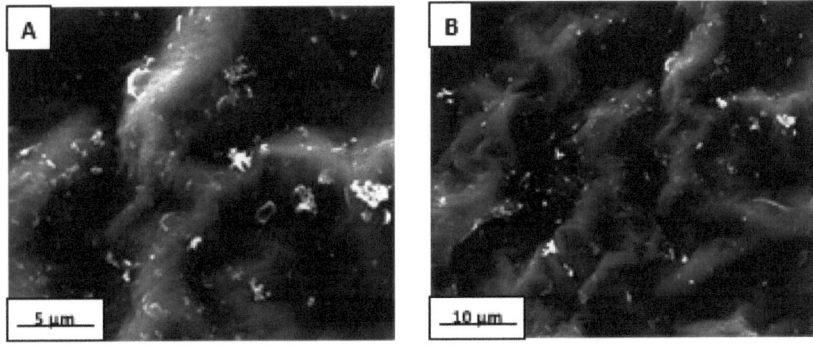

Figure 12. SEM images for A, B) RGO – (EGCG Pt), sheets.

The HI acid and EGCG show a unique ability in reducing PtCl42- to Pt NPs and then depositing these nanoparticles on the surface of sheets. Images from a morphological study conducted by SEM are depicted in Figure 13. All the electroless deposition samples show a good distribution, small size particles and less aggregation with EGCG treatment, unlike the samples that were treated with only HI acid without using EGCG.

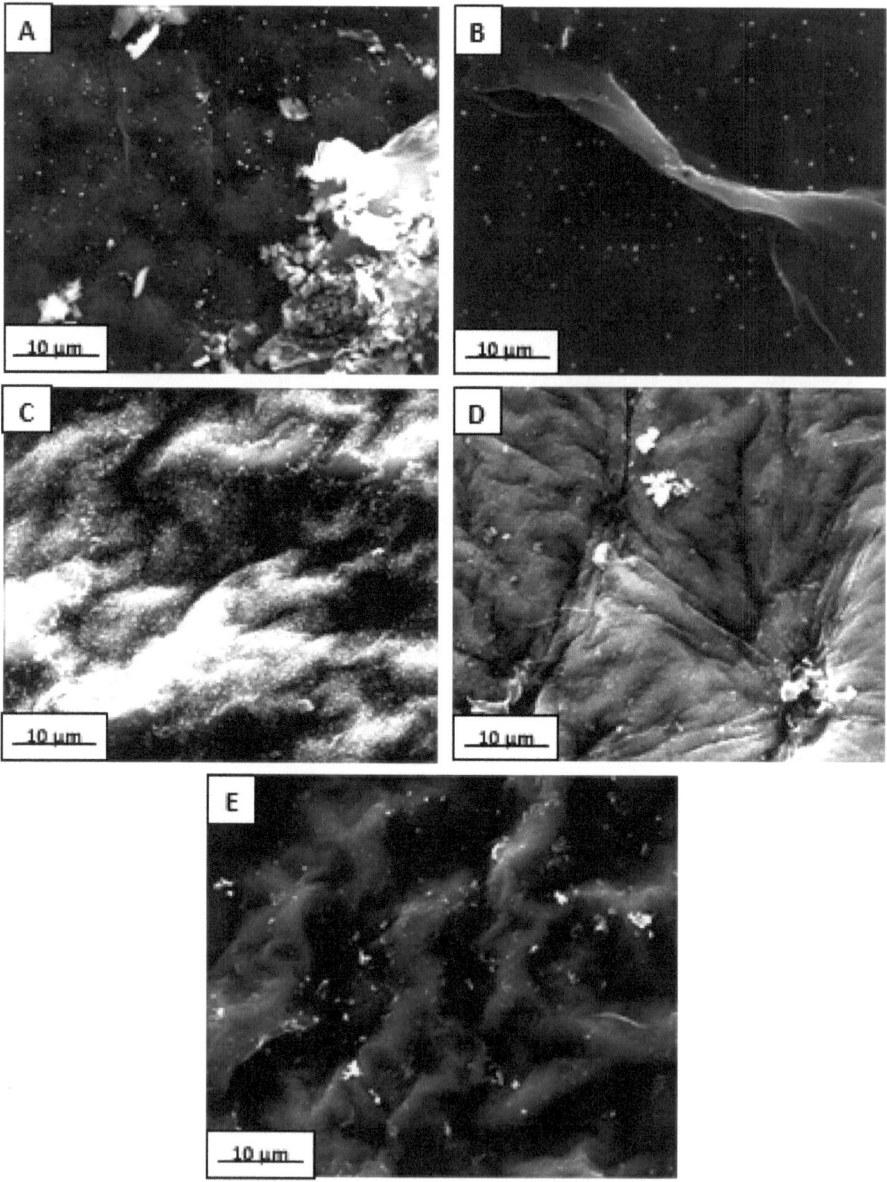

Figure 13. SEM images for A) (rGO – EGCG) – Pt, B) (rGO – 2EGCG) – Pt, C) RGO – (Pt) at 5 hours, D) RGO – (Pt) at 24 hours, E) RGO – (EGCG Pt), sheets.

Electrochemical Activity of Pt Nanoparticles on the surface of different sheets:

The electrochemical behaviour and catalytic activity of all electroless deposition catalysts, prepared with different loading of Pt, has been studied in this section. The cyclic voltammetry (CV) for all electrode sheets has been investigated in 0.1 M H_2SO_4 electrolyte at a scan rate of 100 mV/s.

Figure 14 displays the cyclic voltammograms of GO, GO – (EGCG Pt), (rGO – 2EGCG) and (GO – 2EGCG) – Pt electrodes at a scan rate of 100mV/s. (rGO–2EGCG) – Pt sheet exhibited a much higher current density than that of the pure (rGO – 2EGCG), GO, or GO – (EGCG Pt). The GO – (EGCG Pt) sheet shows a slightly higher current density when compared with the pure GO sheet. The sheet of GO – (EGCG Pt) still has a very low conductivity and a very high resistance, because this sheet has the lowest Pt NPs loading, as well as low reduction of oxygen functional groups.

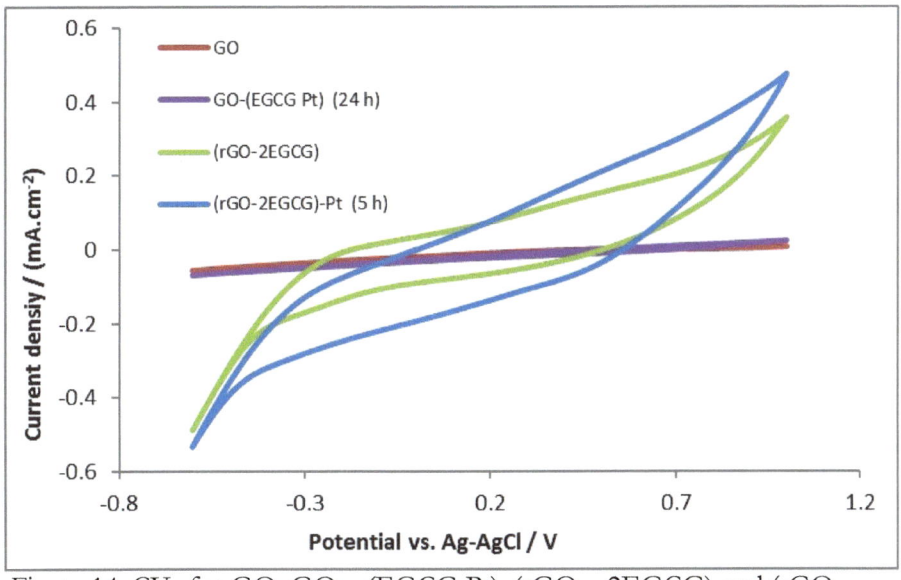

Figure 14. CVs for GO, GO – (EGCG Pt), (rGO – 2EGCG) and (rGO – 2EGCG) – Pt, sheets in 0.1 M H2SO4 solutions at a Scan Rates of 100 (mV/s).

In Figures 15 and 16, the RGO – (Pt) sheet reduced for 24 hour exhibited a much higher current density than all other samples with or without Pt NPs. This signifies that the RGO sheet has a larger surface area and more nucleation sites as a substrate. The CV curve of RGO – (EGCG Pt) which is obtained from treating the RGO sheet with the solution of (Pt salt + EGCG) has a similar result to the SEM and EDX results. After a chemical deposition of RGO sheet for 24 hours with the Pt salt solution, an additional

Pt oxidation peak (~ 0.7 V) and Pt reduction peak (~ 0.4 V) appeared. All results of the samples tested by CV agreed with the SEM's and EDX's analysis results.

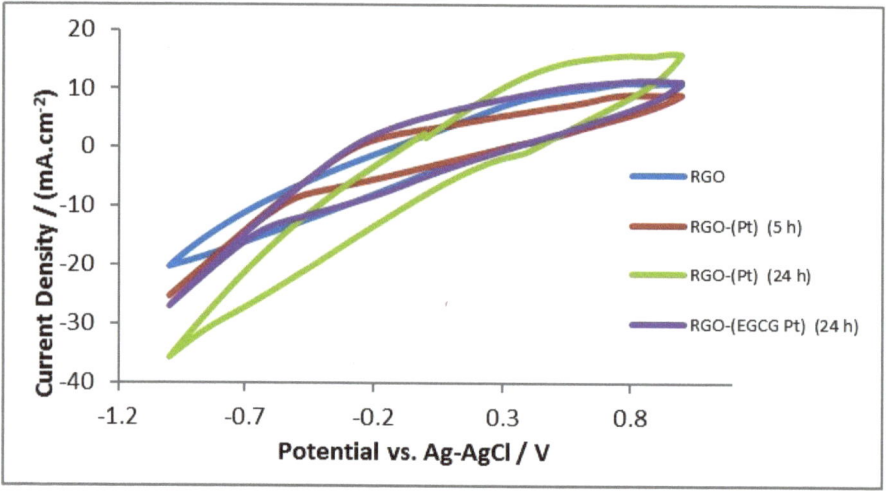

Figure 15. CVs for RGO, RGO – (Pt) reduced in 5 and 24 hours and RGO – (EGCG Pt), sheets in 0.1 M H2SO4 solutions at a Scan Rates of 100 (mV/s).

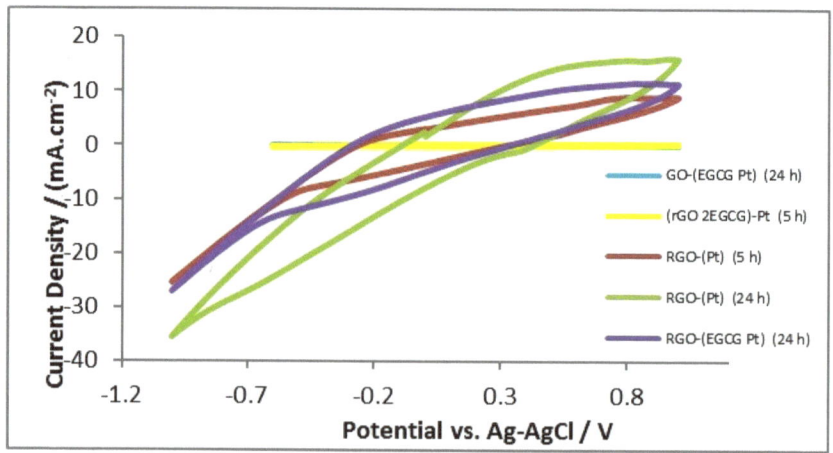

Figure 16. CVs for GO – (EGCG Pt), (rGO – 2EGCG) – Pt, RGO – (Pt) reduced in 5 and 24 hours,   RGO – (EGCG Pt) for 24 hours, sheets in 0.1 M H2SO4 solution at a Scan Rates of 100 (mV/s).

3.3- Synthesis and characterization of Pd electrochemical deposition on Reduced GO, to form graphene - Pd nanoparticles compositions:

"Electroplating deposition" is the application of a metal coating to a metallic or other conducting surface by simple electrolysis of an aqueous solution containing the desired complex in the electrochemical process. Electrodeposition is a simple, fast, green technique, inexpensive, highly stable and reproducible method for preparation of Pd nanoparticles. This process involves using a two- or three-electrode cell system, where the electrolyte generally serves as the source of Pd as well as the conductive medium [14]. By changing the conditions of electrochemical deposition the size and shape of nanoparticles are controlled. The potential or current density of an electrochemical cell can be controlled to deposit electrochemically [14]. There are a variety of electrodeposition techniques that we can use to fabricate Pd and Pd-based nanomaterials, such as chronopotentiometry (CP). CP is a technique for electrodeposition that is carried out by applying a constant current or a current step and the resulting potential change is plotted versus time.

Electrodeposition of PdNPs from acidic electrolyte at RGO sheet electrode; (RGO – Ed Pd) sheet:

Many papers reported the difficulty in producing stable Pd deposition from a high acidic electrolyte bath without any additional reduction agent at a pH less than 5, especially if the base is a non-metal substrate like graphene oxide [20]. In the present work, the Pd nanoparticles were deposited on the surface of RGO from acidic electrolyte bath compositions based on (5 gm/L PdCl2 + 0.1 M HCl) with (pH ~ 1 – 1.5) for 1 and 5 hours at a constant current of 6.2 mA.cm-2. This electrodeposition work successfully fabricated a very high Pd loading electrode without the aid of any reduction agent by a one-step electrochemical co-deposition approach. The anionic complex forms PdCl4-2 in the presence of HCl according to the following equation [5]:

$$Pd^{+2} \ + \ 4\,Cl^- \ + \ 2\,H^+ \ \longleftrightarrow \ 2H^+ \ + \ PdCl_4{}^{2-}$$

The electrodeposition of Pd from PdCl42- in HCl media as an aqueous complex of Pd(II) was reported in many works to be [20]:

$$PdCl_4{}^{2-}\,(aq) \ + \ 2\,e^- \ \longrightarrow \ Pd(s) \ + \ 4\,Cl^-\,(aq) \qquad E^0 = 0.62 \ V$$

The uniform coating of Pd nanoparticles on the surface of RGO was shown in SEM images (Figure 17 and 18) with the average particle size ~ 4 nm diameter. These nanoparticles grow into three-dimensional dendritic structures, rising from a nonlinear diffusion process on the surface of the electrode.

Moreover, the size and density can be tuned by changing the reaction time and current density of the deposition process. Multiple layers of Pd were built up by only increasing the time of electrodeposition to 5 hours with the same electrolyte and technique, Figure 18.

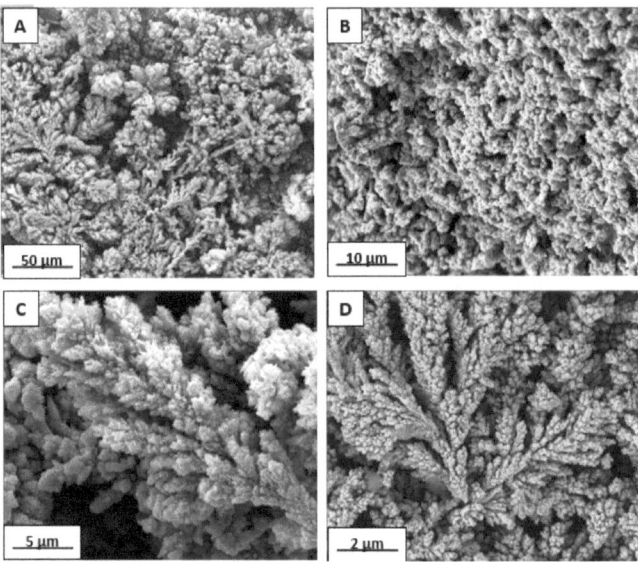

Figure 17. SEM images for (RGO – Ed Pd) sheet after 1 hour palladium electrodeposition from PdCl2+HCl electrolyte.

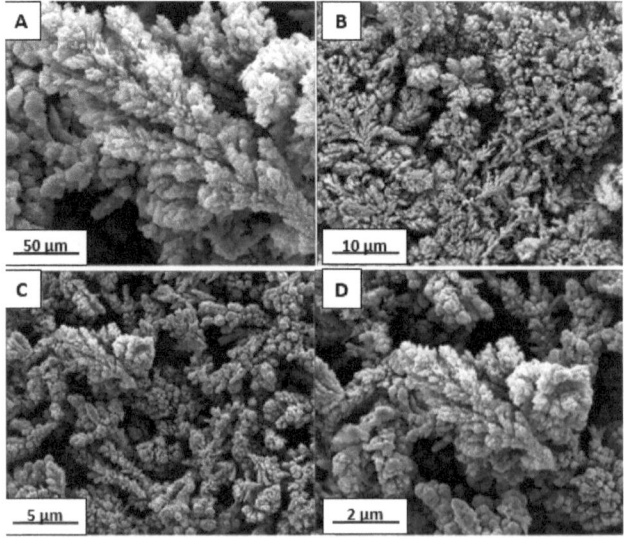

Figure 18. SEM images for (RGO –Ed Pd) sheet after 5 hours palladium electrodeposition from PdCl2+HCl electrolyte.

24

The chemical and structure of RGO – Ed Pd deposited for 1 and 5 hours were further studied with EDX spectroscopy element mapping. The EDX analysis verifies the presence of C, O, I, Cl and Pd elements in the RGO – Ed Pd sheet electrode. Figures 19 and 20 show the SEM images and corresponding EDX spectrum, and the chemical maps demonstrate that all five elements are homogeneously distributed. In Figure 19, the predominant EDX peak corresponding to Pd is attributed to the success of Pd electrodeposition process by recording Pd (~ 91.3 wt %) loading for just 1 hour of electrodeposition.

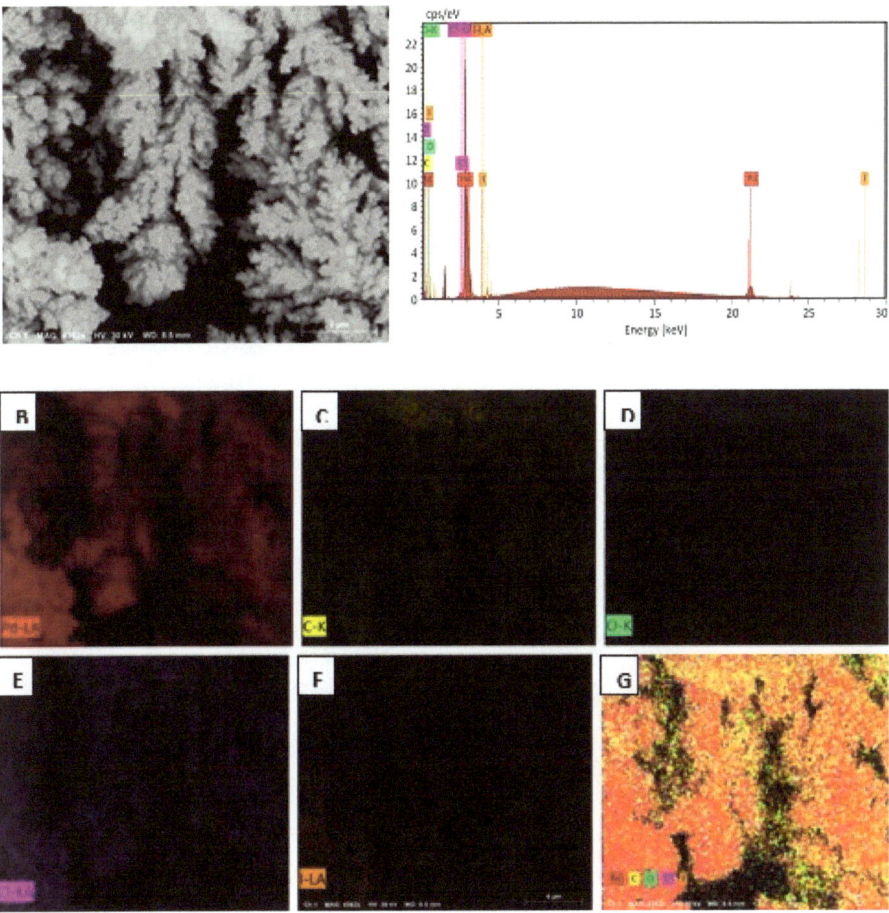

Figure 19. A, B, C, D, E, F, G) SEM image and corresponding elemental mapping of RGO – Ed Pd electrodeposition for 1 hour ; H) EDX spectrum of the selected area in (A).

The major peaks in EDX analysis presented in Figure 20 corresponded to Pd (more than 98 wt %) loading, confirming the almost complete coating RGO surface electrode with Pd NPs. Furthermore, the mapping in Figure 20 shows a high dispersal of Pd NPs on the surface after 5 hour deposition.

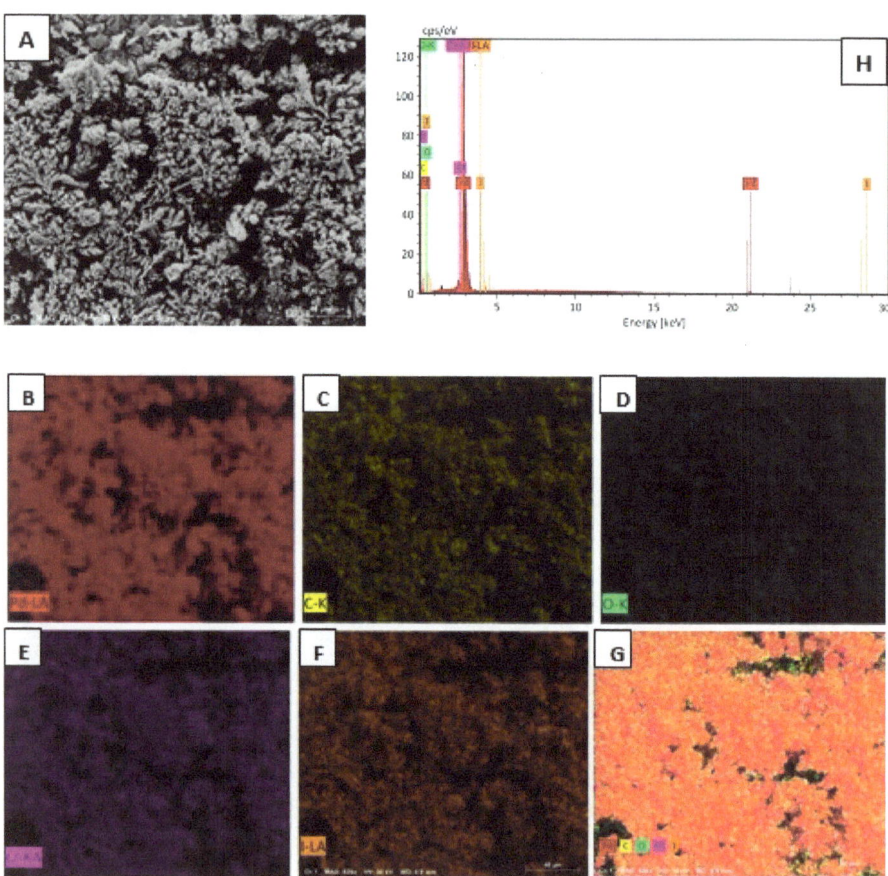

Figure 20. A, B, C, D, E, F, G) SEM image and corresponding elemental mapping of RGO – Ed Pd electrodeposition for 5 hours; H) EDX spectrum of the selected area in (A).

Electrocatalytic performance of RGO – Ed Pd sheet:

To evaluate the catalytic activity of electrochemical deposition catalyst after preparing RGO – Ed Pd with two different periods loading of Pd NPs, we have studied these samples with cyclic voltammetry (CV). All electrode sheets were investigated in 0.1 M H_2SO_4 electrolyte at a different scan rate.

Clearly, the CV curves exhibit three traditional distinctive potential regions in the potential range of (1.2 to -1) V. In general, Figures 21 and 22 show that the shape of voltammograms for both 1 and 5 hours Pd electrodeposition are almost the same. For 5 hour electrodeposition, peak A at ~0.5V is related to Pd reduction, peak B is realted to hydrogen adsorption and evolution. An anodic peak at ~0.83 V is realted to hydrogen desorption overlayed with Pd oxidation peak. Figure 21 shows that by increasing the electrochemical deposition time the Pd reduction peak shifted to more negative potentials. The results also show that current density related to hydrogen formation and evolution increased by inceasing Pd electrodeposition time on the surface of RGO. In Figure 22, the CV curves of the modified RGO with Pd (RGO – Ed Pd) sheet show a significant increase in the electroactive properties compared to RGO.

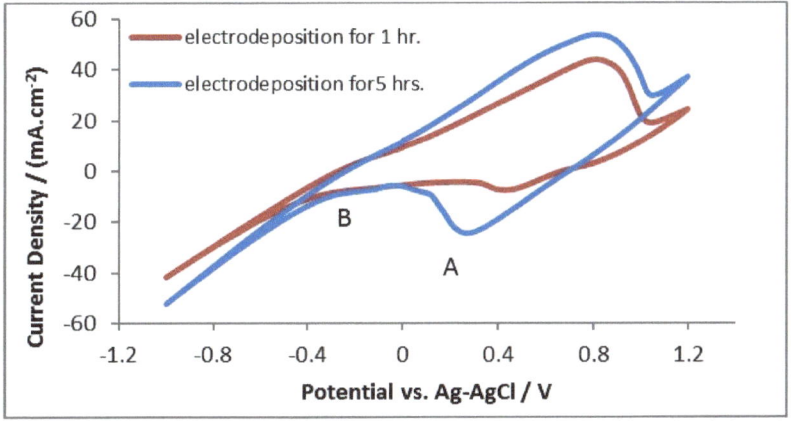

Figure 21. CV curves of RGO – Ed Pd sheet electrodeposited for 1hour (red) and 5hours (Blue), in 0.1 M H2SO4 solutions at the scan rates of 20 (mV/s).

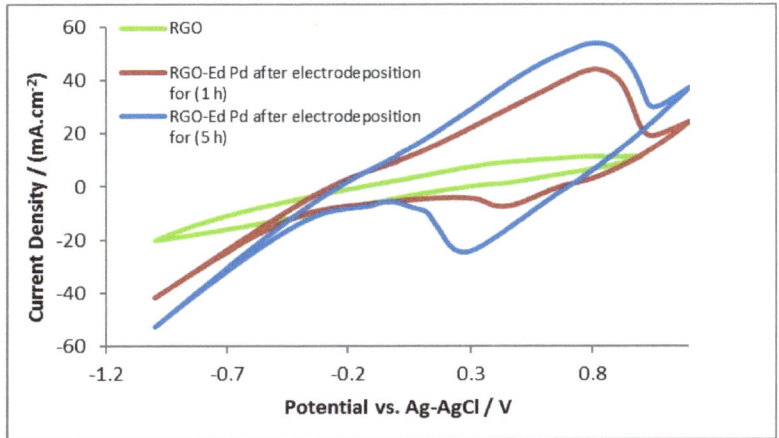

Figure 22. CV curves of RGO and RGO – Ed Pd sheet electrodeposited for (1, 5) hours in 0.1 M H2SO4 solutions at Scan Rates of 20 (mV/s).

Figures 23 and 24 show CVs of Pd – RGO electrodes at different scan rates from 10 to 100 mV/s. The current density of the hydrogen oxidation peak increased and shifted towards more positive potentials by increasing the scan rate. The Pd and hydrogen reduction peaks increased and shifted to more negative potentials with increasing scan rates.

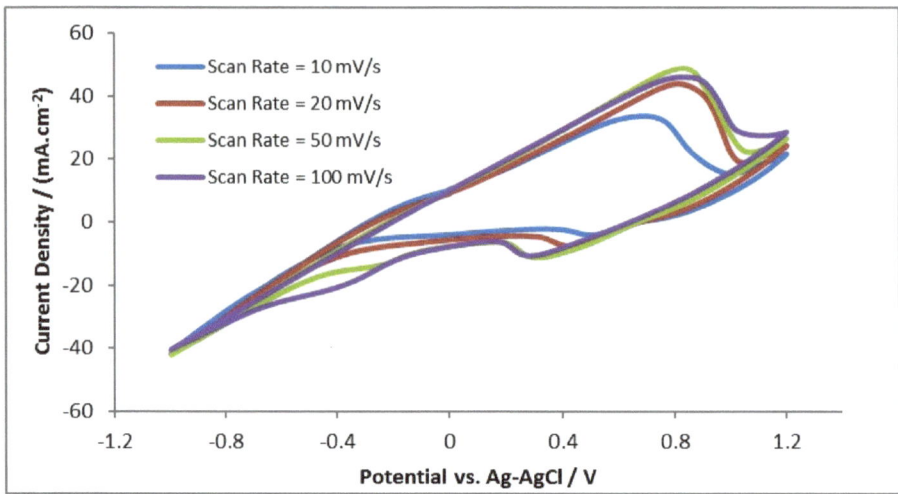

Figure 23. CVs curves of RGO – Ed Pd sheet (1 hour) electrodeposition in 0.1 M H2SO4 solutions at different Scan Rates of (10, 20, 50, 100) (mV/s).

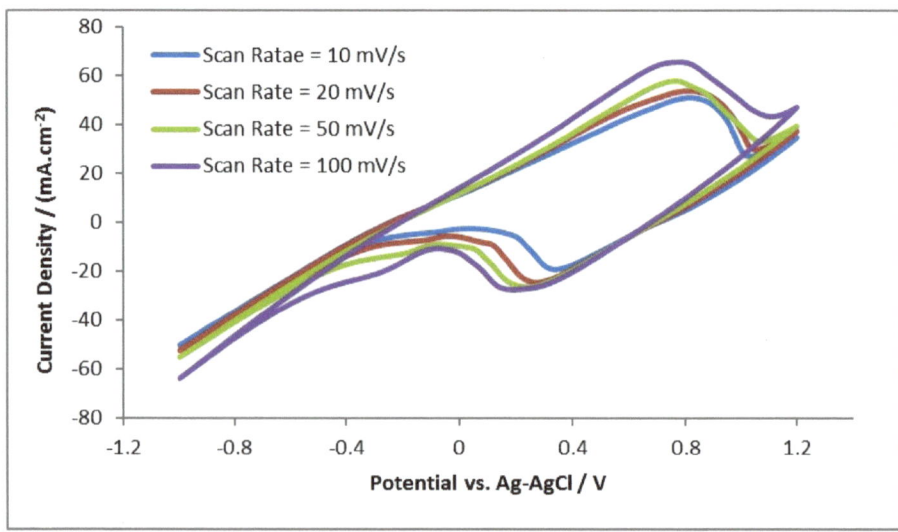

Figure 24. CVs curves of RGO – Pd sheet (5 hours deposition) in 0.1 M H2SO4 solutions at different scan rates of 10, 20, 50, 100 mV/s.

In Figures 25 and 26, the RGO – Ed Pd sheet electrode was examined by increasing the higher potential limit. The results reveal that the Pd reduction peak increased by increasing the higher potential limit, and shifted to more negative potentials which is related to the formation of a higher concentration of Pd oxide on the surface at a higher positive potential scan range.

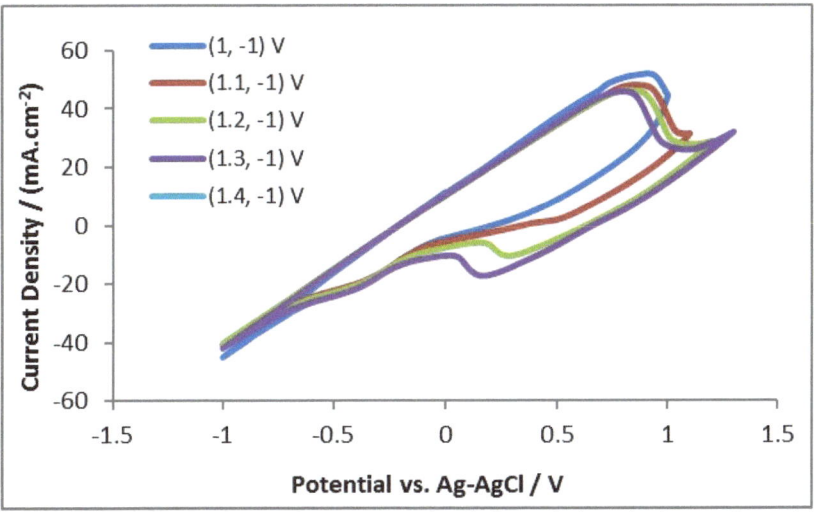

Figure 25. CVs curves of RGO – Ed Pd sheet (1 hour) electrodeposition as a function of upper potential windows for a fixed lower potential of -1V, in 0.1 M H2SO4 solutions at Scan Rates of 100 mV/s.

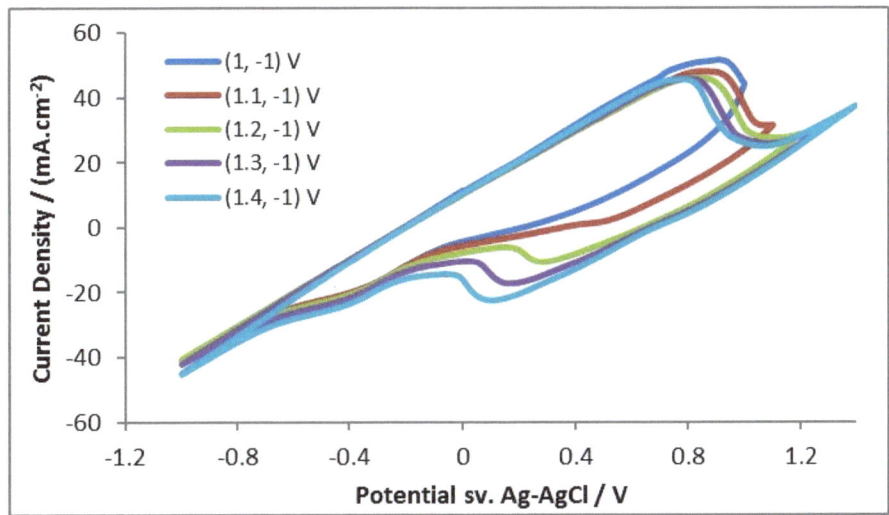

Figure 26. CVs curves of RGO – Ed Pd sheet (5 hours) electrodeposition as a function of higher potential windows for a fixed lower potential of -1V, in 0.1 M H2SO4 solutions at Scan Rates of 100 mV/s.

Electrodeposition of Pd NPs from Pd nanoparticles suspension with EGCG at the surface of RGO sheet electrode (alkaline bath); (RGO – Ed Pd NP):

Much effect in electroplating from precursor solution has been conductor to improve excellent properties for electrocatalytic production. There have been many efforts to change the electrodeposition conditions in order to obtain smooth deposition, complete coverage with a small-sized particle, control of the shape of particles and high efficiency of electrodeposition. This electrodeposition can be controlled by the bath temperature, pH, and operating current density [20]. Most chemical formulations that are used to deposit Pd electrochemically are neutral to alkaline in order to improve the stability and achieve the desired morphologies of Pd NPs on the surface of the electrode. This work successfully produced an effective electrocatalyst with a high covering of nanoparticles on the surface of the electrode with a higher efficiency, stability and less amount of Pd without changing any electrodeposition conditions by using water base Pd NP suspension. For this experiment, we tried to deposit Pd electrochemically at room temperature from 0.1 M LiOH + Pd NP suspension on the surface of RGO sheet by using the same technique in the previous electrodeposition work for acidic electrolyte bath. The reduction behavior of EGCG has been used to prepare a stable strong alkaline ($\sim$ 14 pH) aqueous solution of Pd NP suspension [21]. To prepare RGO – Ed Pd NP sheet electrode, electrodeposition was carried out in an alkaline Pd NP aqueous solution for two different deposition times, 24 and 48 hours at a constant current of 6.2 mA.cm-2. As depicted in SEM in Figures 27 and 28, The Pd NP was successfully deposited on the surface of the RGO sheet in a spherical shape with even distribution and many pores.

In Figures 27 and 28, the Pd NP started the formation of small dendritic structures on the surface of the electrode then built up multiple layers with increasing electrodeposition time from 24 hours to 48 hours.

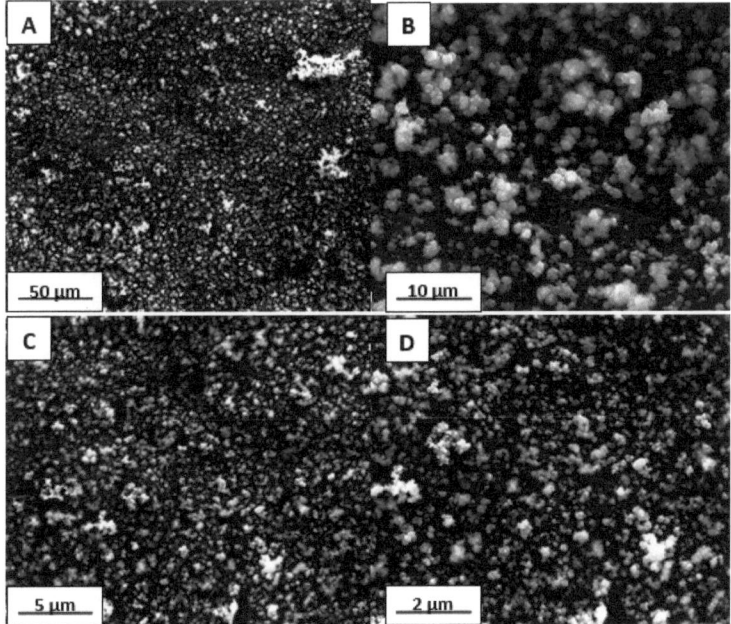

Figure 27. SEM images for (RGO – Ed Pd NP) sheet after 24 hours palladium electrodeposition from Pd NP + LiOH electrolyte.

Figure 28. SEM images for (RGO – Ed Pd NP) sheet after 48 hours palladium electrodeposition from Pd NP + LiOH electrolyte.

The EDX elemental mapping pattern in Figures 29 and 30 show the Pd NPs with a very high uniform distribution. Additionally, EDX spectrum analysis for composition elements shows the major peaks corresponded to Pd NP (~ 53.65 wt %) loading for 24 hours and (~ 91.28 wt %) loading for 48 hours.

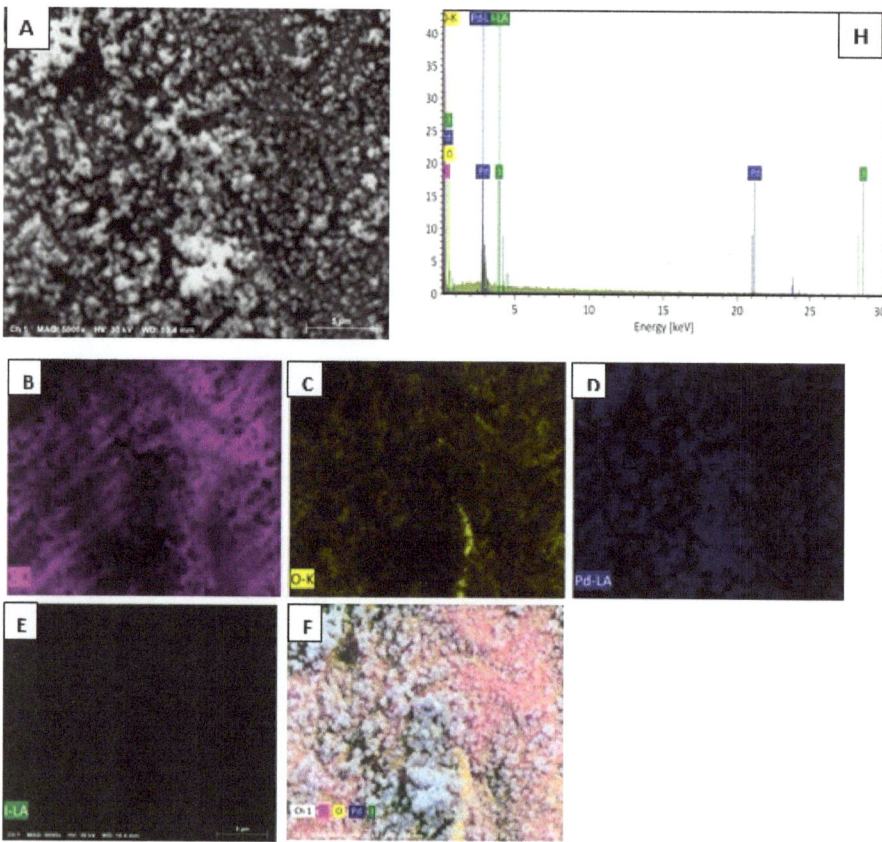

Figure 29. A, B, C, D, E, F, G) SEM image and corresponding elemental mapping of RGO – Ed Pd NP electrodeposition for 24 hours; H) EDX spectrum of the selected area in (A).

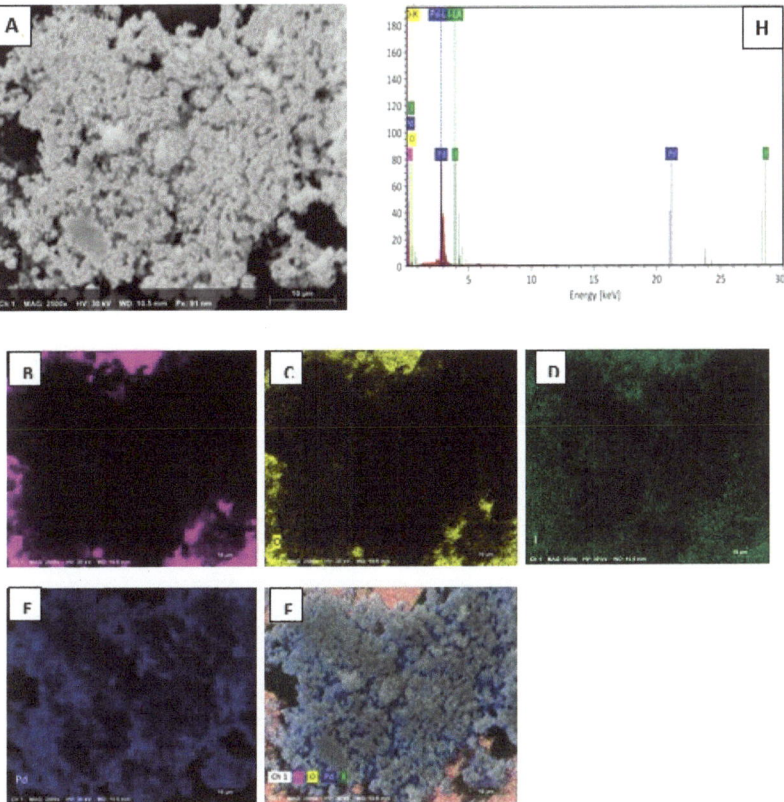

Figure 30. A, B, C, D, E, F, G) SEM image and corresponding elemental mapping of RGO – Ed Pd NP electrodeposition for 48 hours; H) EDX spectrum of the selected area in (A).

The average particle size of deposition Pd on the surface of RGO sheet is ~ 2 nm diameter. In the preparation for Pd NP suspension, the Pd was encapsulated by EGCG.

In this electrolyte, EGCG acted as a stabilizer because of the strong capping effect through the chemical interaction of the carbonyl groups or phenolic hydroxyls with Pd NPs [10].

Electrocatalytic performance of RGO – Ed Pd NP electrode sheet: To evaluate the catalytic ability of electrochemical deposition after preparing RGO – Ed Pd NP with two different loading periods of Pd NP, we have studied these samples with cyclic voltammetry (CV). All electrode sheets are investigated in 0.1 M H2SO4 electrolyte at a different scan rate.

In Figure 31, the CV curves in the potential range of (1.2 to -1) V show additional peaks which were exhibited after electrodeposition of Pd NP. For 24 hours of electrodeposition, peak A at ~ 0.32 V is related to Pd reduction, peak B is realted to hydrogen adsorption and evolution, an anodic peak at ~ 0.48 V is related to hydrogen desorption overlayed with Pd oxidation peak. In comparing with a RGO sheet electrode, the RGO – Ed Pd NP electrode has the higher current density especially after doubling the electrodeposition time from 24 hours to 48 hours. This result is caused by the high concentration loading of Pd NP on the surface of RGO electrode which leads to an increase the active surface area of the RGO – Ed Pd NP sheet. The hydrogen oxidation peak at ~ - 0.173 V is higher after 48 hours of electrodeposition compared to 24 hours electrodeposition because of more loading of Pd NP on the surface of the electrode and as a result more hydrogen adsorbtion and absorption into Pd. The CV curve after electrodeposition of Pd NP on RGO for 24 hours shows the hydrogen redction peak at ~ - 0.196 V which is clearer than the same peak that appeared on the CV after 48 hours of electrodeposition, because the latter overlaped with the hydrogen evalution region.

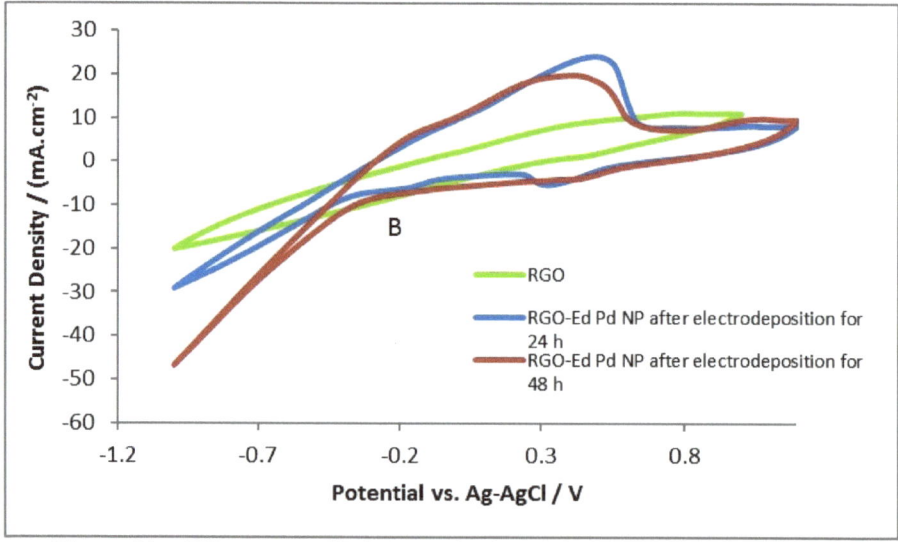

Figure 31. CV curves of RGO– Ed Pd NP sheet deposited for (24, 48) h in 0.1 M H2SO4 solutions at Scan Rates of 20 (mV/s).

Figures 32 and 33 show CVs of EGCG Pd – RGO electrodes at different scan rates from 10 to 200 mV/s. The current density of the hydrogen oxidation peak increased and shifted towards more positive potentials by increasing the scan rate. The Pd and hydrogen reduction peaks increased and shifted to more negative potentials with increasing scan rates.

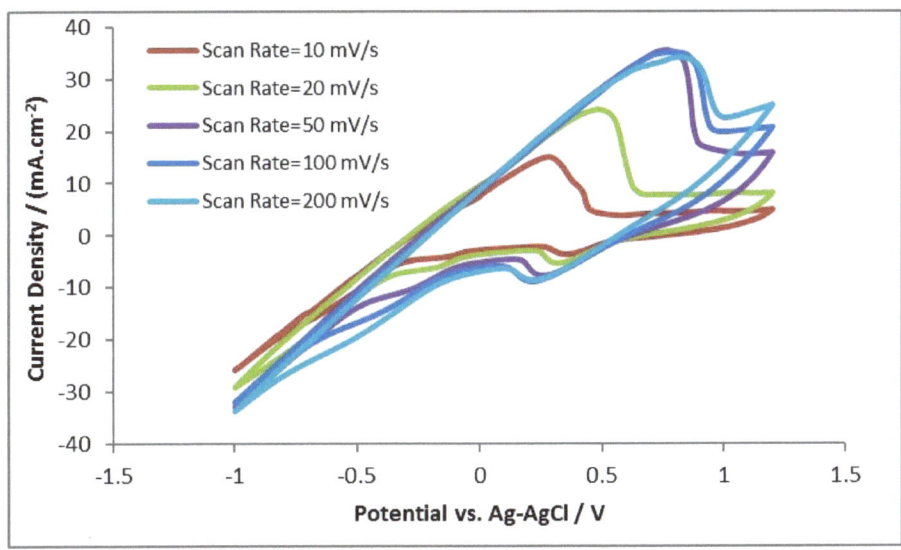

Figure 32. CV curves of RGO– Ed Pd NP sheet (24 hours deposition) in 0.1 M H2SO4 solutions at different Scan Rates of (10, 20, 50, 100, 200) (mV/s).

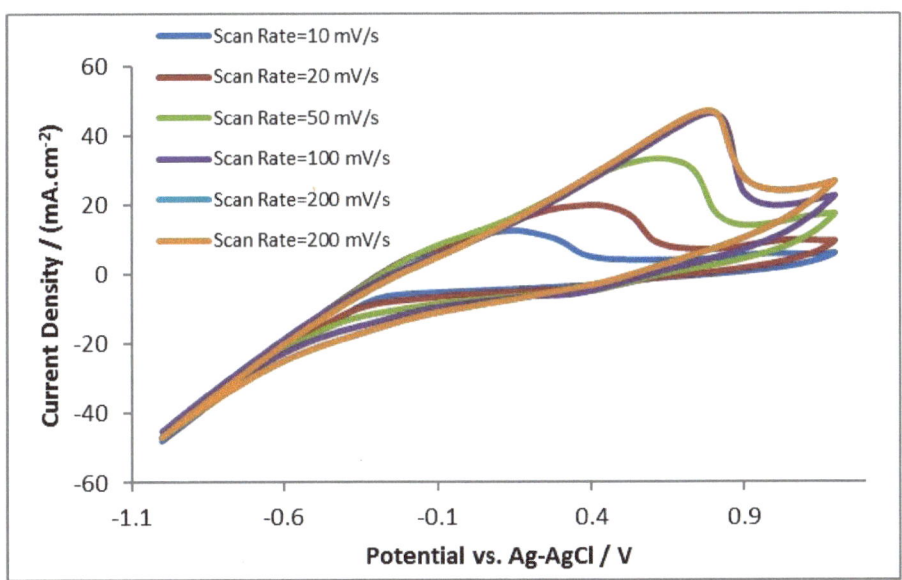

Figure 33. CV curves of RGO – Ed Pd NP sheet (48 hours deposition) in 0.1 M H2SO4 solutions at different Scan Rates of (10, 20, 50, 100, 200) (mV/s).

In Figures 34 and 35, the RGO – Ed Pd NP sheet electrode was examined by increasing the higher potential limit. The results reveal that the Pd reduction peak increased by increasing the higher potential limit, and shifted to more negative potentials which is related to the formation of a higher concentration of Pd oxide on the surface at a higher positive potential scan range.

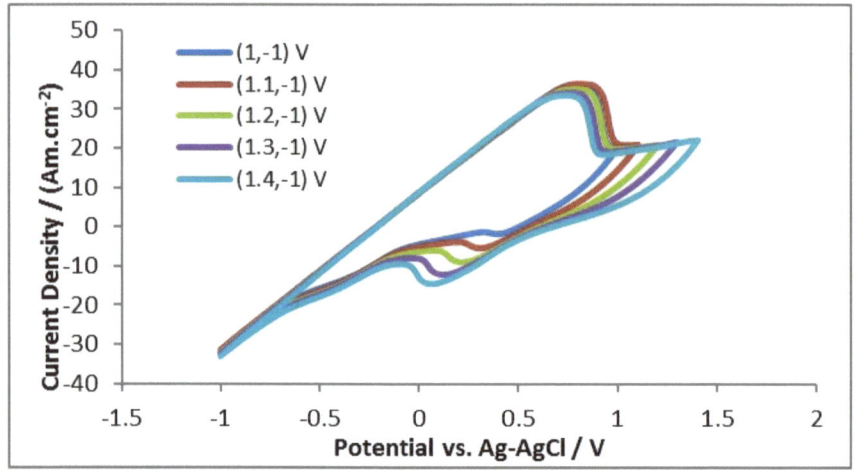

Figure 34. CV curves of RGO – Ed Pd NP sheet (24 hours electrodeposition) as a function of upper potential windows for a fixed lower potential of -1V, in 0.1 M H2SO4 solutions at Scan Rates of 100 mV/s.

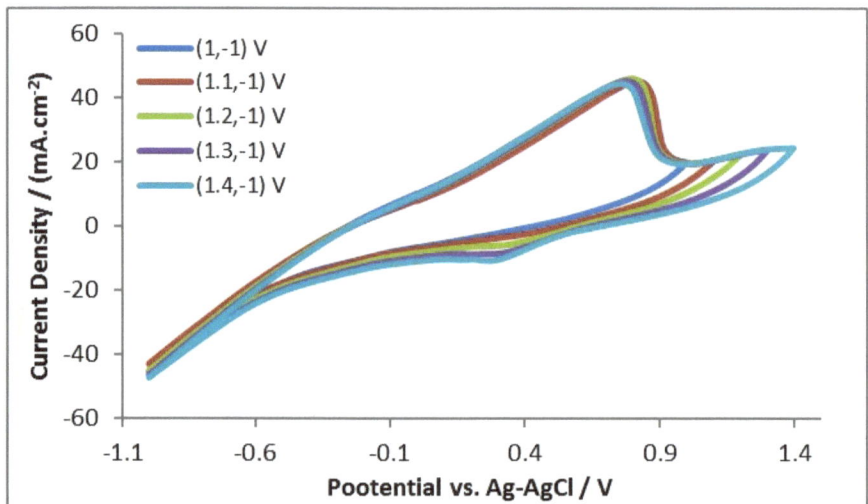

Figure 35. CV curves of RGO – EGCG Pd sheet (48 hours electrodeposition) as a function of upper potential windows for a fixed lower potential of -1V, in 0.1 M H2SO4 solutions at Scan Rates of 100 mV/s.

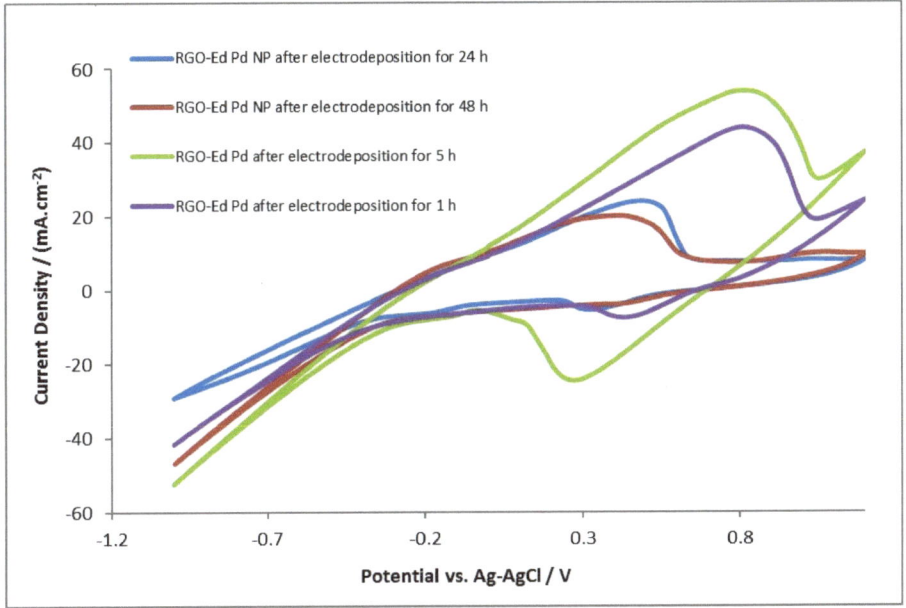

Figure 36. CV curves of RGO– Ed Pd sheet electrodes deposited for (1, 5) hours and RGO – Ed Pd NP RGO sheet electrodes deposited for (24, 48) hours in 0.1 M H2SO4 solutions at Scan Rates of 20 (mV/s).

The effect of two different bath compositions, PdCl4-2 + HCl (PdII complex) and 0.1M LiOH + (Pd NP), on Pd electrodeposition on RGO and their catalytic activity were studied using cyclic voltammograms. The CV voltammograms in Figure 36 illustrates that in the whole potential range the current density on

RGO – Ed Pd (prepared from PdII complex bath) is higher than RGO – Ed Pd NP (prepared from Pd NP Suspension) at different reduction periods. In addition, the RGO – Ed Pd shows a higher current density for the hydrogen evolution reaction region. The Pd reduction peak for RGO – Ed Pd is more significant and slightly more shifted to a negative direction than the Pd reduction peak on

RGO – Ed Pd NP sheet.

These results are related to the higher amount of Pd which is loaded on the surface of RGO electrode from PdII complex bath in comparison to RGO modified by palladium nanoparticles from Pd NP suspension. The lower concentration of Pd NP on the surface of RGO from Pd NP suspension is related to the strong capping effect of EGCG. The long time of electrodeposition process with Pd NP electrolyte could be caused by the high capping of these nanoparticles by EGCG and also related to the difficult deposition of nanoparticles. However this capping effect increased the stability of the RGO – Ed Pd NP as a catalyst with a uniform distribution and a smaller average particle size of Pd NP.

The EGCG was used to control the reduction ability of the precursor solution with a strong stabilizing effect of forming Pd NP [14]. The cross sectional area in Figure 37 of RGO – Ed Pd NP sheet shows that there was almost no immobility of Pd NP onto the RGO sheet again due to strong capping effect of EGCG.

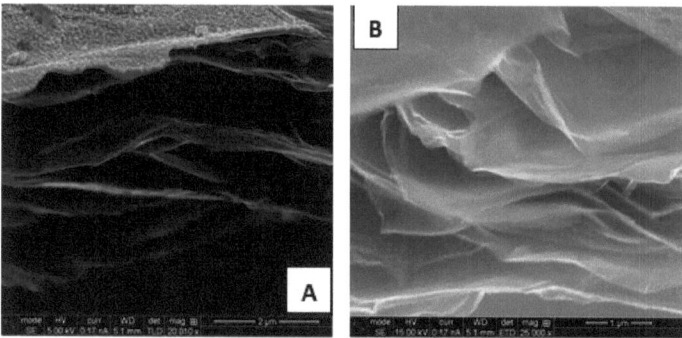

Figure 37. SEM images of A and B) the cross-section of the RGO – Ed Pd NP sheet.

In spite of getting more Pd NP loading on the surface of RGO using PdII complex bath, the results show that the RGO modified with Pd NP from Pd NP suspension has a higher stability with a uniform coverage and a smaller particle size, which is more favorable in electrocatalytic applications. The results reveal that to compare the catalytic activity of these two catalysts more accurately we need to control the total amount of Pd at the surface of both catalysts and keep them at the same level.

Table 2: Samples name and their brief preparation procedure.

No.	Sample	Product	Procedure
1	Graphene oxide sheet	GO	GO sheet modified by "Hummers and Offenman's" method.
2	GO sheet in platinum salt solution	GO – (Pt)	GO sheet immersed in Pt salt solution.
3	GO sheet in EGCG solution	GO – (EGCG)	GO sheet immersed in EGCG solution.
4	GO sheet in the solution of (EGCG & Pt salt)	GO – (EGCG Pt)	GO sheet immersed in the solution of (EGCG & Pt salt).
5	Reduced GO by EGCG in liquid media	rGO – EGCG	GO & EGCG suspension →dried → sheet of rGO – EGCG
6	R`GO – EGCG sheet in Pt salt solution	(rGO – EGCG) – Pt	rGO – EGCG sheet immersed in Pt salt solution.
7	Reduced GO by two times EGCG in liquid media	rGO – 2EGCG	GO & 2 times of EGCG suspension →dried → sheet of rGO – EGCG
8	R`GO – 2EGCG sheet in Pt salt solution	(rGO – 2EGCG) – Pt	rGO – 2EGCG sheet immersed in Pt salt solution.
9	Reduced GO sheet by HI acid	RGO	GO sheet immersed in HI acid for1 h.
10	RGO – HI sheet in Pt salt solution	RGO – (Pt)	RGO sheet immersed in Pt salt solution.
11	RGO – HI sheet in EGCG solution	RGO – (EGCG)	RGO sheet immersed in EGCG solution
12	RGO – HI sheet in the solution of (EGCG & Pt salt)	RGO – (EGCG Pt)	RGO sheet immersed in the solution of (EGCG & Pt salt).
13	R`GO – 2EGCG sheet reduced further by HI acid	R(rGO – 2EGCG) – HI	rGO – 2EGCG sheet immersed in HI acid.
14	Electrodeposition of PdNPs on the surface of RGO – HI sheet	RGO-Ed Pd	Pd electrodeposited on the surface of RGO sheet from $PdCl_2$ acidic bath.
15	Electrodeposition of PdNPs on the surface of RGO sheet using PdNp +EGCG suspension	RGO- Ed Pd NP	EGCG Pd electrodeposited on the surface of RGO sheet from PdNp +EGCG suspension alkaline bath.

CHAPTER 4
CONCLUSION

In summary, at first we reduced graphene oxide (GO) sheet which was used as a substrate for electrocatalysis in our work by HI acid and EGCG reduction agents. Comparing the results from reduction work by using HI acid and EGCG reduction agents, the electrical conductivity for RGO sheet which is obtained from HI acid treatment is higher and shows a very good flexibility with a much higher tensile strength than the samples with the presence of EGCG.

After that, for finding a high active electrocatalysis we used the chemical deposition method to deposit Pt NPs from Pt salt on the surface of different substrates which are obtained from the reduction of GO using HI acid and EGCG. The HI acid and EGCG show a unique ability in reducing $PtCl_4^{2-}$ to Pt NPs and then depositing these nanoparticles on the surface of sheets. All the electroless deposition samples show a good distribution, small sized particles and less aggregation with EGCG treatment, unlike the samples that were treated with HI acid without using EGCG.

As well as, we reported two successful kinds of electrocatlysis by using the electrodeposition method. In this electrodeposition work, we successfully fabricated a very high Pd loading electrode as a RGO – Ed Pd electrode from high acidic electrolyte (PdII complex) bath (~ 1.5 pH) without the aid of any reduction agent by a one-step electrochemical co-deposition approach.

Moreover, for the other electrodeposition work we successfully produced an effective electrocatalyst with a high covering of nanoparticles on the surface of the electrode with a higher efficiency, stability and less amount of Pd without changing any electrodeposition conditions by using water base Pd NP suspension as the electrodeposition electrolyte.

Finally, we compared the RGO – Ed Pd electrode, which is fabricated from PdII complex bath, and the RGO – Ed Pd NP electrode, which is fabricated from Pd NP suspension bath. The latter has a higher stability with a uniform coverage and smaller particle size, which is more favorable in electrocatalytic applications.

REFERENCES

1- Bell, Alexis T. "The impact of nanoscience on heterogeneous catalysis."Science 299.5613 (2003): 1688-1691.

2- Toh, Shaw Yong, et al. "Graphene production via electrochemical reduction of graphene oxide: synthesis and characterisation." Chemical Engineering Journal 251 (2014): 422-434.

3- Mohan, Velram Balaji, et al. "Characterisation of reduced graphene oxide: Effects of reduction variables on electrical conductivity." Materials Science and Engineering: B 193 (2015): 49-60.

4- Liao, Ruijuan, et al. "Polyphenol-reduced graphene oxide: mechanism and derivatization." The Journal of Physical Chemistry C 115.42 (2011): 20740-20746.

5- Rao, Chepuri RK, and D. C. Trivedi. "Chemical and electrochemical depositions of platinum group metals and their applications." Coordination Chemistry Reviews 249.5 (2005): 613-631.

6- Hummers Jr, William S., and Richard E. Offeman. "Preparation of graphitic oxide." Journal of the American Chemical Society 80.6 (1958): 1339-1339.

7- Xiong, Dongbin, et al. "Oxygen-containing functional groups enhancing electrochemical performance of porous reduced graphene oxide cathode in lithium ion batteries." Electrochimica Acta 174 (2015): 762-769.

8- Wang, Yan, ZiXing Shi, and Jie Yin. "Facile synthesis of soluble graphene via a green reduction of graphene oxide in tea solution and its biocomposites." ACS applied materials & interfaces 3.4 (2011): 1127-1133.

9- Pei, Songfeng, and Hui-Ming Cheng. "The reduction of graphene oxide."Carbon 50.9 (2012): 3210-3228.

10- Pei, Songfeng, et al. "Direct reduction of graphene oxide films into highly conductive and flexible graphene films by hydrohalic acids." Carbon 48.15 (2010): 4466-4474.

11- Sun, Yimin, et al. "Real-time electrochemical detection of hydrogen peroxide secretion in live cells by Pt nanoparticles decorated graphene–carbon nanotube hybrid paper electrode." Biosensors and Bioelectronics 68 (2015): 358-364.

12- Sun, Li-ming, Chen-lu Zhang, and Ping Li. "Characterization, antimicrobial activity, and mechanism of a high-performance (−)-epigallocatechin-3-gallate (EGCG)− CuII/polyvinyl alcohol (PVA) nanofibrous membrane." Journal of agricultural and food chemistry 59.9 (2011): 5087-5092.

13- Esmaeilifar, A., et al. "Synthesis methods of low-Pt-loading electrocatalysts for proton exchange membrane fuel cell systems." Energy 35.9 (2010): 3941-3957.

14- Xiao, Lisong, et al. "Enhanced in vitro and in vivo cellular imaging with green tea coated water-soluble iron oxide nanocrystals." ACS applied materials & interfaces 7.12 (2015): 6530-6540.

15- Tyagi, Deepak, Salil Varma, and S. R. Bharadwaj. "Pt/graphite catalyst for hydrogen generation by HI decomposition reaction in S − I thermochemical cycle." International Journal of Energy Research 39.15 (2015): 2008-2018.

16- Ilda, Itsuo. "The kinetic behaviour of the decomposition of hydrogen iodide on the surface of platinum." Zeitschrift für Physikalische Chemie 109.2 (1978): 221-232.

17- Macagno, V. A., M. C. Giordano, and A. J. Arvia. "Kinetics and mechanisms of electrochemical reactions on platinum with solutions of iodine-sodium iodide in acetonitrile." Electrochimica Acta 14.4 (1969): 335-357.

18- Shukla, Ravi, et al. "Laminin receptor specific therapeutic gold nanoparticles (198AuNP-EGCg) show efficacy in treating prostate cancer." Proceedings of the National Academy of Sciences 109.31 (2012): 12426-12431.

19- Chen, Aicheng, and Cassandra Ostrom. "Palladium-Based Nanomaterials: Synthesis and Electrochemical Applications." Chemical Reviews 115.21 (2015): 11999-12044.

20- Schlesinger, Mordechay, and Milan Paunovic, eds. Modern electroplating. Vol. 55. John Wiley & Sons, 2011.

21- Private communication with Professor Dr. Kattesh Katti's.

ABOUT THE AUTHOR

Nada works at the Iraqi Ministry of Oil. She is a Chemical Engineer who had her Master's Degree in Science from the University of Missouri-Colombia in 2016 as part of the HCED scholarship granted by the Prime Minister's Office in Iraq. She worked in her MA project with professor John Gahl at University of Missouri-Colombia.

www.ingramcontent.com/pod-product-compliance
Lightning Source LLC
Chambersburg PA
CBHW040921180526
45159CB00002BA/566